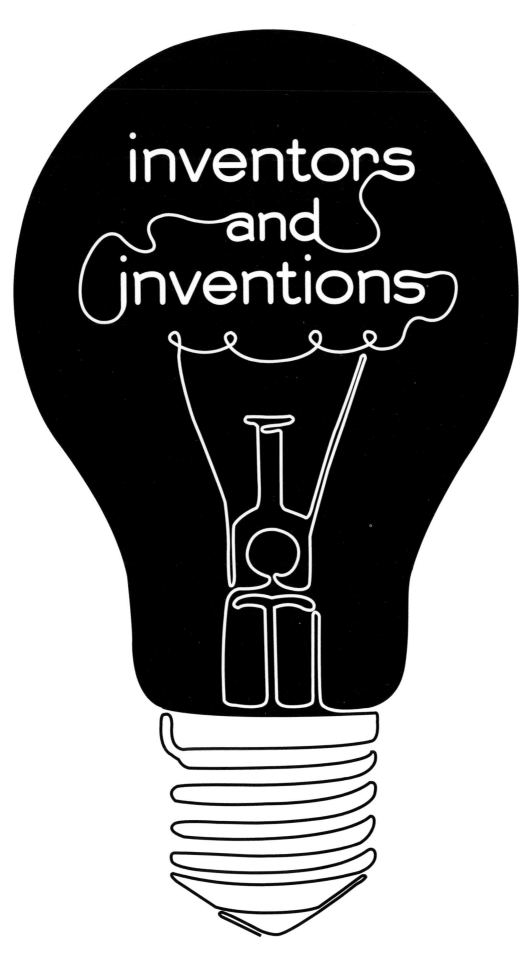

inventors
and
inventions

black dog
publishing
london uk

CONTENTS

AN INTRODUCTION TO INVENTION

06   **Richard Fisher**

EARLY INVENTIONS

12   **Time**
14   **The Numbering System**
16   **Money**
18   **The Alphabet**
20   **The Wheel**

DOMESTIC

24   **Light Bulb**
26   **Pressure Cooker**
27   **Aga Cooker**
28   **Microwave Oven**
29   **Ready Meal/TV Dinner**
30   **Pop-up Toaster**
32   **Refrigeration**
32   **Tetra Pak**
34   **Washing Machine**
35   **Dish Washer**
36   **Vacuum Cleaner**
38   **Central Heating**
39   **Air-conditioning**
40   **Soap**
41   **Shower**
42   **The Flush Toilet**
43   **Toilet Paper**
44   **Toothbrush/Toothpaste**
46   **Safety Razor**
47   **Scissors**
48   **Perfume**
49   **Aerosol**
50   **Sewing Machine**
52   **Safety Match**
53   **Umbrella**
54   **The Zip**

ENTERTAINMENT

58   **Photography**
60   **Television**
64   **Cinema**
68   **Gramophone**
70   **Cassette Tape**
71   **Personal Stereo**
72   **Compact Disc**
73   **MP3**
74   **Scrabble**
75   **Monopoly**
76   **Chess**
78   **Lego**
79   **Rubik's Cube**
80   **Soduku**
81   **Crossword Puzzle**
82   **Yo-Yo**
84   **Frisbee**
85   **Hula Hoop**
86   **Ice Skates**
87   **Roller Skates**
88   **Skateboard**
89   **Roller Coaster**
90   **Electric Guitar**
91   **Moog Synthesizer**
92   **Slinky**
94   **View Master**
95   **Etch A Sketch**
96   **Kaleidoscope**
98   **Chewing Gum**

COMMUNICATION

102   **Pencil**
103   **Pencil Sharpener**
104   **Eraser**
105   **Quill Pen**
106   **Fountain Pen**
107   **Ballpoint Pen**
108   **Gutenberg Press**
110   **Typewriter**
112   **Paper**
114   **Linotype Composing Machine**
115   **Fax Machine**
116   **Computers**
118   **Internet**
119   **Radio Broadcasting**
120   **Phonograph**
122   **The Postal Service**
124   **Telephone**
126   **Global Positioning System (GPS)**
127   **Radar**

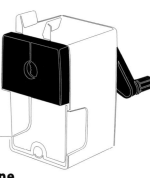

ENGINEERING AND TRANSPORT

130 **Archimedes Screw**
132 **Ball Bearing**
134 **Universal Joint**
135 **The Dynamo**
136 **Bicycle**
138 **Motorcycle**
139 **Segway HT**
140 **Motorcar**
142 **The Hybrid**
144 **Roads and Tarmac**
146 **Traffic Lights**
147 **Cats Eyes**
148 **Hovercraft**
150 **Trains and Railways**
154 **Ornithopter**
155 **Hot Air Balloon**
156 **Aeroplane**
160 **Jet Engine**
162 **Rocket Propulsion**
164 **The Skyscraper**
165 **Circular Saw**
166 **Window Pane**
168 **Cement**
170 **Synthetics**

MEDICINE

174 **Vaccination**
175 **Band-Aid**
176 **Spectacles**
178 **Hearing Aid**
179 **Stethoscope**
180 **Scanning**
181 **Iron Lung**
182 **Artificial Heart**
183 **Pacemaker**
184 **Prosthetics**
185 **False Teeth**
186 **Viagra**
187 **Contraception**
188 **In Vitro Fertilisation (IVF)**
189 **Aspirin**
190 **Genetic Engineering**
191 **Cloning**

WARFARE

194 **Bows of War**
195 **Catapult**
196 **Hand Guns**
198 **Machine Gun**
200 **Tanks**
202 **Land Mines**
203 **Swiss Army Knife**
204 **Dynamite**
205 **Hand Grenade**
206 **Atomic Bomb**
208 **Submarine**
209 **Night Vision**

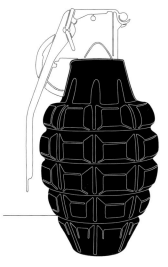

EXPLORATION

212 **Maps**
214 **Telescope**
216 **Compass**
217 **Astrolabe**
218 **Barometer**
219 **Gyroscope**

AGRICULTURE AND FOOD

222 **Food Preservation**
224 **Tractor**
226 **Plough**
227 **Combine Harvester**
228 **Irrigation**
229 **Rotary Tiller**
230 **Pesticides**
231 **Fertilisers**
232 **Animal Husbandry**
233 **Windmill**
234 **Gristmill and Waterwheel**

Richard Fisher

AN INTRODUCTION
TO INVENTION

Remove the back of your mobile phone and you are likely to find an unassuming object hidden inside that has driven a revolution in electronics in recent decades. The Lithium-ion battery is not exactly an exciting invention; you are unlikely to see technology's first-adopters writing enthusiastic blogs about its energy-to-mass ratio. But Li-ion batteries provide the power for everything from the iPod to the laptop, and eventually they may even sit inside your car instead of an internal combustion engine. The batteries are continually finding applications in the latest pieces of technology, which call on their ability to pump out much more energy from their tiny, lightweight cells than their weaker counterparts such as nickel metal hydride cells. The Li-ion battery also happens to demonstrate what invention is really all about.

The concept of the Li-ion battery is pretty straightforward: positively charged lithium ions inside the battery jump between an anode and a cathode. This process releases electrons into an external circuit that delivers the power for your device. Then, when you plug the battery in to recharge, all of the lithium ions shift back to the anode, which becomes replete with atoms of lithium so that the process can be repeated. The idea that this chemical reaction could be used in rechargeable cells was proposed by Michael Stanley Whittingham, then at Exxon, in the 1970s. But his invention was far from the finished product. It took many more years of research on the chemistry of layered oxides before Sony could release the first fully commercialised Li-ion battery, which went on to revolutionise the consumer electronics industry.

Right Varta Lithium-ion battery.

Opposite Tesla Roadster electric powered sports car.

hen Whittingham made the first rechargeable Li-ion prototype, he used a cathode of titanium sulfide and lithium metal as the anode. One of the big problems with this design was that a lithium metal anode posed severe safety issues, since it was harder to keep the materials stable. It was not until the early 1980s that a team at Bell Labs managed to produce a safe and workable battery using an anode made from graphite, and it would take much longer for the battery to be made ready for commercial use.

And failures in the battery's design have continued. It seemed faults in the batteries were allowing metal particles to clog up the separator between the anode and the cathode. Most of the time, this would cause the battery to power down, but in rare instances, it could cause overheating and fire. A couple of years ago, alarming pictures of a battery inside a laptop exploding and catching fire at a conference in Japan circulated on the web. The company that manufactured the laptop recalled around four million of its computers fitted with the battery.

Such setbacks have been a key driving force for inventors to come up with new ideas. For example, in 2008 researchers at the Fraunhofer Institute for Silicate Research ISC in Wurzburg, Germany, announced that they had developed a prototype battery with electrodes made from non-flammable material, unlike the current electrodes that burn much more easily. No doubt this year will see many more improvements in the design of the battery—if it is not superseded by something altogether different.

So what does the Li-ion battery's history tell us about invention? First, an obvious point: an invention is about a lot more than just its inventor. How we apply it, improve it, tweak it, rebuild it and refashion often defines it just as much as its inventor, and doing so is often the key to its success and longevity. All inventions continually evolve, getting better, cheaper and faster. If they don't, they are superseded, and the Li-ion battery is no different. It is often the case that we credit an invention to one person. You can blame the 'light bulb above the head' image for that. Though it is true that human creativity can occasionally generate such a 'eureka' moment, more often than not it requires the work of more than one person to allow an idea to realise its full potential.

The Li-ion battery also demonstrates another facet of invention: the birth and very often the success of an invention is intertwined with what came before and progress that is happening all around. Take the case of electrically powered transportation. Many scientists and companies hope that we will soon ditch the dirty internal-combustion engines inside our cars, and go electric. Among the highest profile of these vehicles currently on the market is the Tesla Roadster, built by a company headed by Elon Musk, the founder of the internet payment service PayPal. The Roadster is an electric dream—with a top speed of 130 mph and the capacity to accelerate from 0–60 mph in under four seconds. It was ranked number two in *Time* magazine's 2008 Inventions of the Year (pipped to the top spot by a retail DNA test kit, which allows people to discover their genetic predispositions to conditions or traits). The car gets its juice from a 450 kilogram Li-ion battery pack rumoured to cost tens of thousands of dollars—nearly a quarter of the cost of the car itself.

ut the fact that the Li-ion power pack will deteriorate over time could be a major showstopper for the car. The cells have a relatively short life span because the battery's cathode tends to wear out. This is something consumers can just about live with inside a mobile phone—they can replace it or just dispense with the phone—but it poses much more of a problem as the devices that the batteries are powering get bigger. It means the distance that the Roadster can drive per charge will shrink—even quicker if you happen to live in a hot region. That means the proud owner of a Tesla will need a hugely expensive replacement within a few years after buying the car. The Tesla owes much of its existence to the Li-ion battery, but its eventual fate is also intertwined with future developments in the technology. We tend to think of inventions individually, but if you could draw the relationships between these ideas, you would see a vast web of mutually reliant links stretching right back through history.

Despite its scope for improvement, the Li-ion battery is a relatively 'finished' product. We rarely hear of the inventions that simply did not work, but it is important that we acknowledge the importance of those that fell by the wayside in invention. It is an unfortunate fact for the individuals involved that many inventors, if not most of them, will fail to translate their idea into a 'success'. It might be down to an unforeseen technical hitch, a competitor pipping them to the post, or perhaps simply because investors run out of patience and capital dries up. You need spend just five minutes browsing patents to see how many ideas never go any further than a blueprint. Yet if failure is not celebrated and catered for in society then ingenuity withers. In his 2008 book *Hot, Flat and Crowded*, the author Thomas Friedman argues that if we want green technologies to combat climate change then we need 10,000 inventors in 10,000 garages and laboratories working on 10,000 crazy ideas. Though nearly all will fail, he points out, perhaps one of them will emerge into the real world with something that changes society. This is the essence of invention: without the incremental steps, wrong turns and failures, the progress of invention stalls.

In fact, the original invention of the battery itself by Alessandro Volta in 1800 was built on another scientist's failed theory. Volta was driven to his invention after becoming locked in a dispute with another Enlightenment scientist: Luigi Galvani had announced that he could produce muscle contractions in dead frogs by placing them in a circuit with dissimilar metals. Galvani's explanation for the phenomenon was 'animal electricity' and he was convinced that the source of the electricity came from within the tissue of the animal itself. Volta strongly disagreed, and built a set of alternating copper and zinc discs separated by brine-soaked fabric to prove Galvani wrong. Sure enough, his 'voltaic pile' showed that animal tissue was not needed to create a current.

Top Illustration depicting Luigi Galvani, whose experiments with frogs legs sought to prove the hypothesis that animal matter conducted electricity.

Bottom Gillette Fusion five-blade razor.

Opposite Circular runway patent, showing a landing and take-off facility with a centre area and two surrounding circular runways.

However, Galvani may have been wrong with his explanation, but he was partly right in his ideas. His research eventually led others to find that nerves do carry electrical impulses—what came to be known as 'bioelectricity', a discovery now credited to him. The point is that inventors and scientists with big ideas are well familiar with setbacks and failures, but they can still end up changing the world with work like Galvani. For another example of this hit-and-miss approach from the history of electricity, you need look no further than Nikola Tesla, who lived from the mid-nineteenth to twentieth century. Among other ideas, he came up with alternating current, which is the form in which electricity is delivered to your home, and also the radio and fluorescent lighting. But Tesla also happens to have come up with a few other concepts that gained less traction, such as the 'teleforce' weapon, which came to be known as his "death ray". He also tried and failed to tap into the 'universal energy' of the air itself.

Looking back recently at the highlights of *New Scientist*, we found many of the important inventions of the past 50 years, from the printed electronic circuit to polypropylene plastic. One of my favourites came in 1972, when we greeted the news that Gillette was about to embark on an expensive advertising campaign promoting its new 'shaving device', a two-blade razor touted to achieve a closer shave than with a single cut. The key innovation by the scientist Norman Welsh came from noticing that a razor blade pulls the hair slightly out of its follicle—so-called "beard hysteresis"—and that adding a second blade chops the hair lower down before it can retreat. The company claimed this postponed the five o'clock shadow by two and a half hours. Who would have thought at the time that this would eventually spur innovation in a razor blade 'arms-race' between companies that led to the five-blade razors we have today.

Equally enjoyable was looking back at many of the inventions we have covered over the years that never made it beyond the blueprint—often not surprisingly. For instance, in 1965 we described the potential of the 'circular' runway to reduce airport size; and in 1969, we showed how underground trains that split down the middle could save time by filtering passengers wanting to disembark from those carrying on to subsequent stations. The history of invention is arguably all the richer with ideas like these.

Thankfully, most of the inventions you will discover in this book did make it beyond the blueprint, and went on to change the world. These are the end products; the successful fruits of hours of labour, and they should be rightfully celebrated. They also all share something in common: they only came to be because of the collective endeavour of invention. Each was built on what came before, and succeeded because of the shared goals of many different people, all attempting to translate their ideas into something tangible. Without this rich history, many of the cutting-edge inventions of today and tomorrow could not be realised—and that goes even for the humble Li-ion battery.

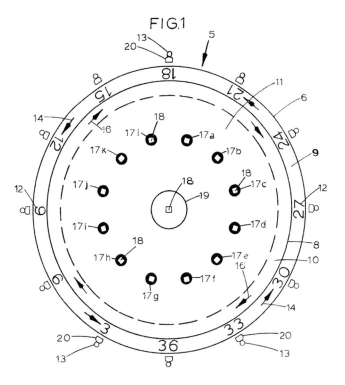

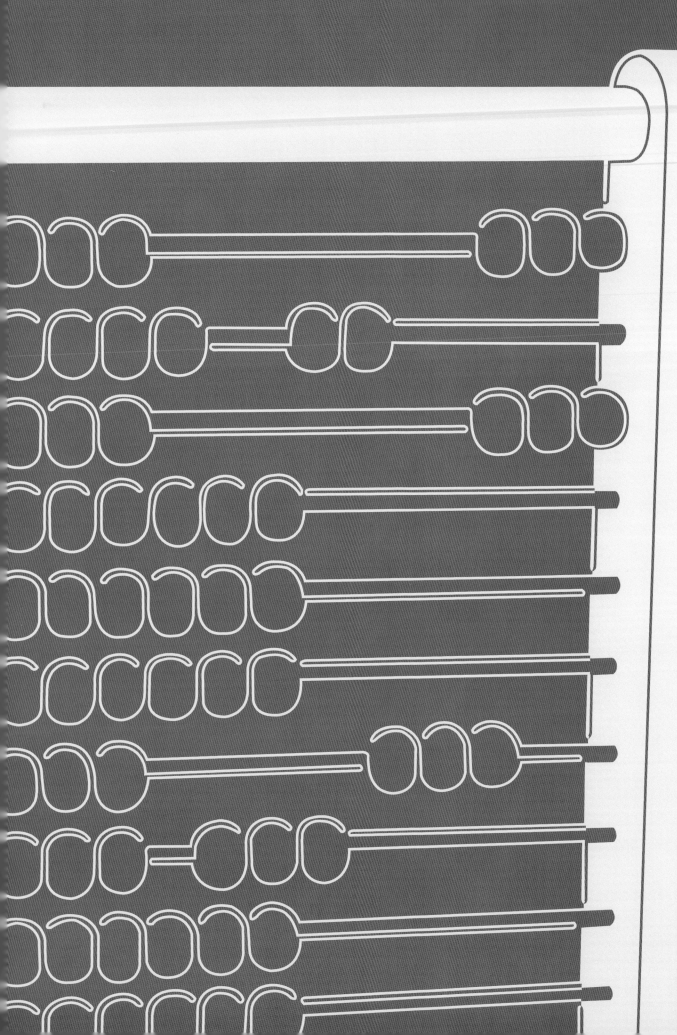

early
inventions

TIME

It might initially seem strange to describe the concept of space and time as a human invention, but the measurement of our universe using these two parameters is, to a large extent an artificial construct. It is in fact observations about changes in our environment—such as the rising of the sun—that have given birth to the documentation of time. The idea of time itself has grown out of the process of documenting movement in space—if nothing moved, time as a concept would make little sense.

The daily rotation of the earth was used as one of the earliest time scales, breaking time up into a series of periods of identifiable and calculable cycles of light and dark. Early Sumerian culture is thought to have used the sun to define time, as did the Romans, whose sundials date back to around 200 BC. However, it does not represent a fundamental scale, as the rotation is not consistent—tidal friction decreases the speed at which the Earth turns, thus changing the length of days.

The other common time divisions similarly look to nature for their lead. The month stems from the period of time taken by the moon to orbit the earth—around 29 days—but this is not strictly adhered to across the world. Rather the month has come to operate, particularly in the West, as simply a convenient time unit by which to subdivide the seasonal year.

The seven-day week, meanwhile, is more arbitrary. In Ancient Greece, the months consisted of three ten-day weeks; whilst the Romans had an eight-day 'market' week that ended in a day of rest and festivals. The USSR, in 1929 and 1932, tried to impose a five-day week, but eventually the familiar seven-day week returned.

The hour was a late development in the history of time, becoming prominent only in the fourteenth century when European towns introduced clocks to their main streets. Before this, more vague systems were in place; such as the Anglo-Saxons' structure of morningtide, noontide and eveningtide.

Minutes were fairly unnecessary throughout most of human history, and it was not until the Industrial Revolution that they really played a role in defining time, particularly in the operation of the workplace. Today, even the minute is not precise enough, and with the increasing speed of operations allowed by the advent of the internet, the second, millisecond and nanosecond have become essential, featuring largely in scientific research.

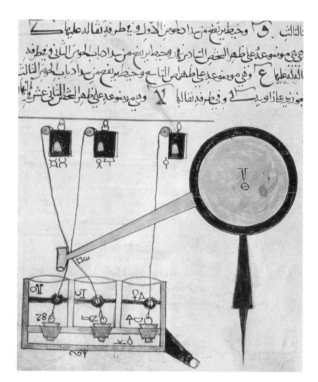

Top and bottom An early clock-making system originating from Istanbul; a typical hourglass.

Opposite top left A rudimentary perpetual calendar showing the placement of days of the week from 1976–2003.

Opposite top right An example of a so-called "calendar coin" produced by James Davies of Birmingham in 1795.

Opposite bottom A sundial including months of the year and relative distances to global cities.

THE NUMBERING SYSTEM

The invention of the number system has developed very slowly over time, with subtle complexities and expansions being introduced in response to practical needs.

It is likely that a primitive numbering system has been in place since the Stone Age. The zero came in later, however, as an interesting device to reflect the absence of a number. The use of negative numbers came in much later still, again through necessity, playing a role in defining loans and debts.

When visualising the number system, it is typical to picture a linear arrangement of numbers of consistently ascending value, starting from zero, with fractional divisions between each whole number. But the picture of the number system as a simple line of regular demarcations is not sufficient. Irrational numbers, such as the endless Pi, or e, have no place on such a regularly defined line. A number such as Pi can never be accurately described, and as such refuses to function as a definable numerical measurement. As a consequence, mathematicians formalised irrational numbers and introduced them amongst the rationals.

Further, equally complex numerical problems have also led to new developments designed to solve them, such as imaginary numbers and hyperreals. The primary reason for adding these new systems has been to try to achieve what is known as a complete system—one that is defined as being able to solve every problem within its own system.

The uncomfortable relationship between the number system and those numbers that refuse to co-operate with it has a long history. The Egyptians and Babylonians were the first to use rational numbers in 3000 BC, but later, in 500 BC, a Greek philosopher sect decided to cover up the troubling discovery of irrationals by murdering the mathematician who discovered them. 200 years later, the truth finally came out at the hands of Euclid.

The Arabic number system was initiated much later, in 600 BC; their system also included the number zero. By the fifteenth century, another complexity was added—the negative number. The next century gave birth to imaginary numbers as a way of devising notational solutions, which in turn led to the development of complex and previously incalculable arithmetic by the seventeenth century.

Top A depiction of the Western Latin numeral system followed by early Babylonian, Chinese, Egyptian and Mayan numerical systems.

Bottom The abacus, an early calculating tool, still used around the world today. Examples of the use of the abacus have been found dating back as far as AD 190.

Opposite Selection of different type faces for today's Western numeral system.

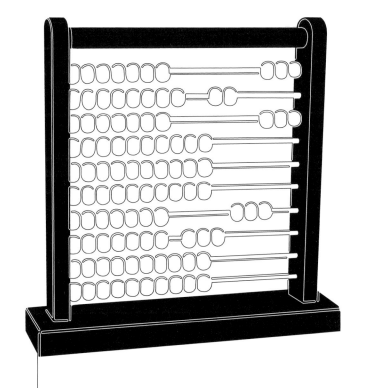

  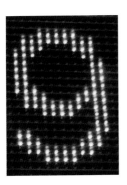

MONEY

Money lies at the very heart of the way most modern societies operate. The use of forms of exchange for the purposes of bartering and trade date back a long way—at least 100,000 years—but early societies without an established monetary system used to rely predominantly on what was known as gift economics—whereby gifts were regularly exchanged without any fixed trade value.

Over time, this somewhat idyllic form of social exchange was replaced with commodity money—that is, an item of certain value that is used to denote that value, such as silver or salt. The shekel was one such early example, used in Mesopotamia around 3000 BC to denote a specific mass of barley. In Asia, Africa, Australia and The Americas, shells were used as tender—usually cowry shells. But it was the Lydians (Lydia is the current day Turkey) who first introduced the gold and silver coin that resembles coinage used today, in around 600 BC.

Commodity money was not necessarily the most practical form of denoting exchange, and over time a more abstracted form of money—representative money—naturally developed. The most direct cause of this was the practice by merchants of issuing receipts to depositors to the value of their deposit—giving rise to a receipt that operated in the same way as money. These developed into bank notes, but continued to exist alongside coinage, and still do. The banknote originated in the Chinese Song dynasty, and did not reach Europe until 1661.

The gold standard, a monetary system where paper notes had a defined value in terms of gold quantities, overtook gold coins as currency during the eighteenth century in Europe. These notes began to be used as tender, and were generally not traded back into gold coinage. By the twentieth century, the gold standard was in wide use across Europe.

Following the Second World War a new dimension arose, whereby most other currencies became fixed to the value of the dollar, which in turn was fixed to gold. However, most currencies became detached again from the US dollar in 1971, when it became no longer possible to convert the dollar back to gold. Now, most currencies rely on the government to attribute value rather than relying on a fixed value commodity.

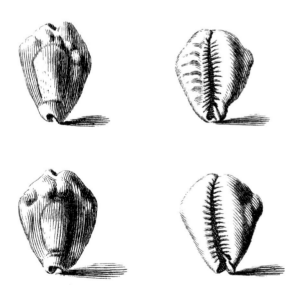

Cowry shells from 1742. Otherwise known as *Cypraea Moneta*. Cowry shells were used as an early form of currency, particularly in the Maldives, Sri Lanka and the Indian Islands.

A B
C D E F G H
I J K L M N
O P Q R S T
U V W X Y Z

THE ALPHABET

The translation of the spoken word into the written word has allowed ideas to be stored, uncorrupted by the inaccuracies of the verbal exchange, across generations and even civilisations. The origins of the alphabet can be traced back to 2700 BC in ancient Egypt, where a system of writing based on 22 hieroglyphs representing different consonants served many functions. However, the glyphs were never actually used purely as a means to document speech. Much later, around 1700 BC, the first 'alphabetic' system is thought to have emerged, although this is open to question. This system does appear to have a strong resemblance to the Egyptian hieroglyphs.

The Proto-Canaanite alphabet eventually developed out of these Egyptian roots, which over time developed further to become the Phoenician alphabet, and in another form as the South Arabian alphabet. All of these scripts however lacked one major element—the vowel.

The first phonemic script was that of the Phoenicians, which contained only about 24 distinct letters, opening it to common use amongst traders. It also could be applied to most languages, as it simply reproduced phonetics.

The Phoenician alphabet spread through the Mediterranean; in Greece, it was modified to include vowels, which represented the first true alphabet. Taking letters that had no attributable Greek sounds, new vowel meanings were then prescribed in their place. The many variants of this consonant-and-vowel based script led to many alphabets with different subtleties. The Latin alphabet developed out of one such strand, and subsequently spread across Europe as the Roman Empire expanded.

After the fall of the empire, the alphabet continued through its use in academic and religious works, eventually evolving into what are now known as the "Romance languages".

Another script, Elder Futhark, is thought to have developed out of the old Italic alphabets, and led to the various Runic alphabets that provided the backbone of the Germanic languages up until the late Middle Ages. Over time, these alphabets have since been replaced by the Latin alphabet, which continues to exist in academia and decorative purposes.

The Cyrillic alphabet, widely used across Eastern Europe and Russia, developed out of the alphabet used to describe the liturgical language Old Church Slavonic, which in its early manifestations was used by ninth century Byzantine Greek missionaries to translate biblical texts, and is still used by some Orthodox and Greek Catholic Churches today.

Language continues to evolve and adapt; the entries in the first credible English dictionary by Samuel Johnson in the eighteenth century already read like a window on a past time, just a couple of hundred years later. Even within a decade, the use of new forms of communication such as mobile phones and emails have led to dramatic new developments in the written word and the way in which modern alphabets are used. The written word refuses to remain static, yet the distant roots of the alphabet remain clearly visible.

Opposite An example of numerous layered typefaces. The basic rudimentary letters are visible in the dense white in the images, with the aesthetic embellishments included on them visible on their peripheries.

Top This copy of the *Duenos Inscription*, found in Rome by Heinrich Dressel in 1880, inscribed on the side of a kernos pot, is one of the earliest recorded example of the Latin alphabet, and the basis for modern Romance-derived examples.

Middle Three signs written in the Cyrillic alphabet. Top to bottom they translate as "Church 5th Century", "Church", and "Exit".

Bottom Egyptian hieratic script, papyrus, circa 1600 BC.

THE WHEEL

Left Leonardo Da Vinci's Worm Gear and Design for the Transmission of Rotary Motion into Alternate Rectilinear Motion.

Top right Diagram depicting the static stability of a wheeled vehicle.

Middle right Leonardo Da Vinci's Treadmill Powered Crossbow, 1485–1488.

The wheel not only exists on almost every form of land-based transport, but also functions throughout modern industry—facilitating movement across the spectrum of contemporary society. Without the wheel, life as it exists today would quite literally come to a standstill. The oldest known example has been identified as originating from Mesopotamia over 5,500 years ago. The actual place and date of its origin is, however, unknown.

The wheel is assumed to have developed from the use of fallen trees or logs as a means to roll items such as large stones in the process of construction. Logs were arranged alongside each other in the form of a sledge, on which the heavy object was placed and then dragged forward. As the object came to the end of the logs, those at the rear would be removed and returned to the front in a cycle, allowing the movement to continue.

Over time, the log rollers that made up these early sledges developed grooves in the centre from repeated use, and it became apparent that with the grooves the process of movement actually became easier. This discovery illustrated a simple law: as the grooved area had a smaller circumference than the unworn area, it used less energy to generate rotation, but the wider 'wheel' area still covered the same distance as before. The grooved area was refined to become the axle, and the wider ends became the wheels.

In time the sledge was developed so that the rolling devices no longer caused the sledge plane and its cargo to roll forward off the rollers. Instead, the axle rotated between pegs fitted into the plane. The next major development was to fix the axle to the sledge so that only the wheels turned, increasing stability and introducing the potential for turning corners.

So effective was the early wheel that its basic design and use has changed little over the centuries. The wheel itself still operates in conveyor belts in a surprisingly similar way to the logs beneath these primitive sledges, whilst the various parts that constituted the early wheel system—such as the axle—remain an integral part of most vehicles.

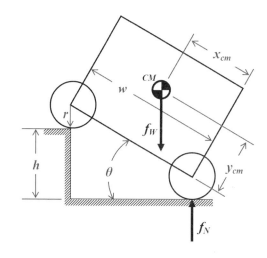

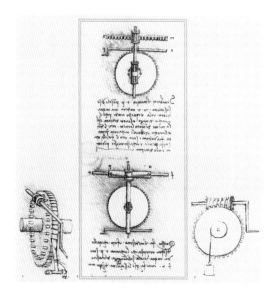

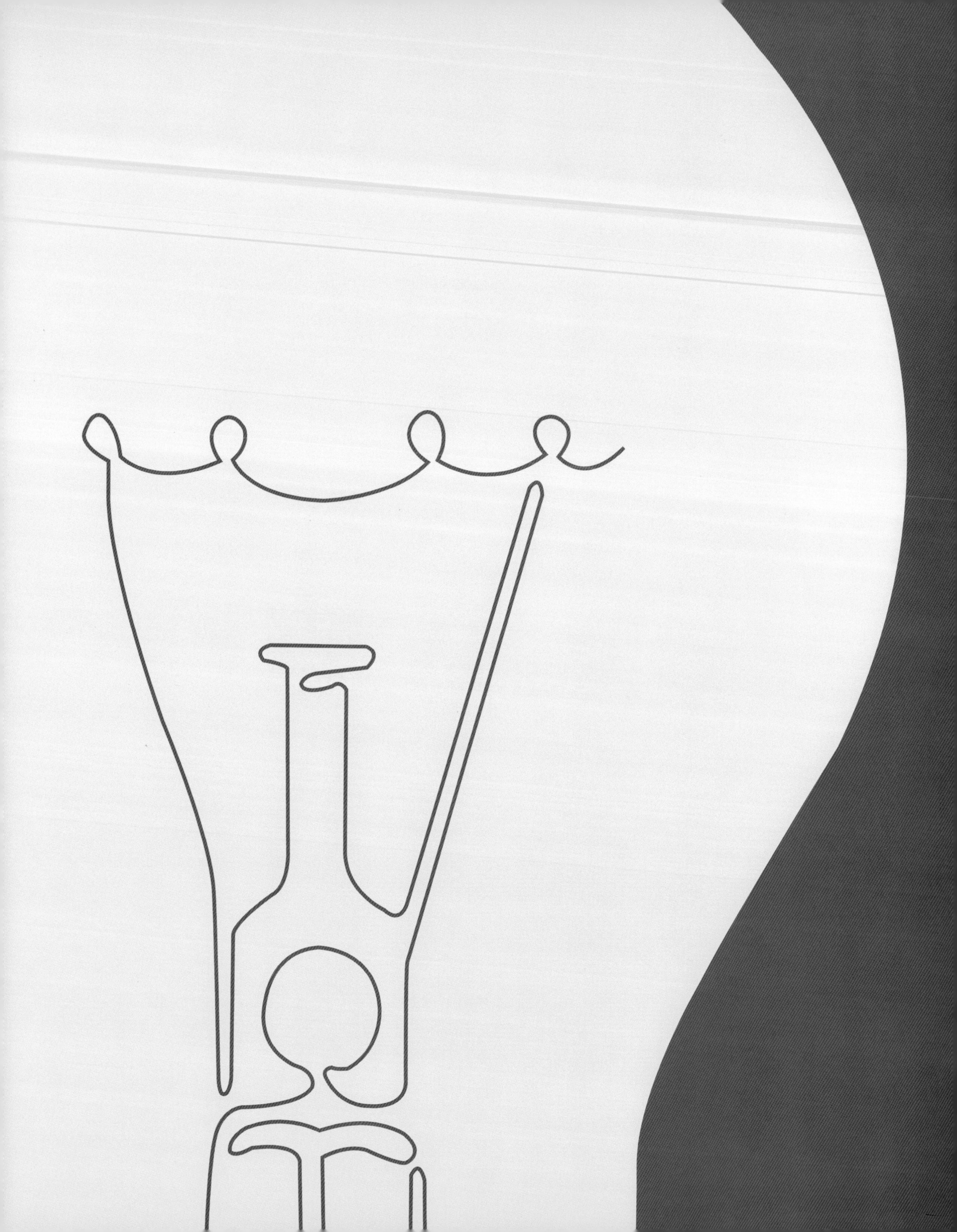

domestic

LIGHT BULB

British chemist Humphry Davy first produced electric incandescent light in 1801 by passing a current through strips of platinum. From this discovery Davy made the first 'carbon arc lamp', which had an electric arc connecting two carbon electrodes. They were reliable but bulky and produced an extremely bright white illumination—far too fierce for domestic use. Its first practical application was in a lighthouse in Dungeness in 1862.

The biggest obstacle encountered by any scientist trying to create electrical incandescent light for domestic use was the speed at which the filaments burnt away. During the late 1840s Joseph Swan's experiments into sustaining this light saw him replace the platinum filaments used by Davy with strips of paper carbonised in tar and treacle, baked in a kiln and then placed within a glass-bulb vacuum. In 1860 he filed the first patent for a light bulb using a carbon filament. However, it still took many more years of research before Swan first demonstrated his light bulb in Newcastle in 1879. The glass vacuum around the filament meant that it kept glowing for longer, but only for a total of 14 hours—not long enough to make Swan's light bulbs a commercial success.

In 1878, a 31 year old Thomas Edison proclaimed that he and his research team at the Electric Light Company would make a safe, cheap and efficient light bulb. After numerous experiments Edison demonstrated the result in October 1879—a bulb with a carbonised cotton thread filament that lasted 150 hours. In 1880, Edison went further and carbonised bamboo fibre to give his light bulb a life of 1,200 hours. If Edison cannot claim to be the inventor of the light bulb he is certainly the innovator of light bulb technology—he filed over 1,092 patents, which provided the foundations for the infrastructure of electric lighting—creating high-voltage generators, transformers, copper-wire power lines and an electric meter. His first installation of a lighting system came in 1880 upon the steamship *Columbia*, along with the erection of the first power plant in Manhattan. Swan was continuing in much the same vein as Edison, illuminating Parliament in 1881 and the British Museum a year later. Eventually the two came to a head regarding patent rights but managed an out-of-court settlement that saw the surprising formation of the Edison and Swan United Electrical Company in 1883. By 1890 many households were enjoying electric light supplied by their very own city power plant.

The next development occurred when tungsten began to replace carbon filaments—a material that has a very high melting point, allowing the filament to shine brighter and for longer. However, there were some drawbacks: tungsten is very hard and brittle and consequently extremely difficult to fashion into thin wire filaments. This problem was overcome by an American named William Coolidge who compressed powdered tungsten into a rod-shaped mould, which was then drawn out into thin wire under extreme heat and pounding. However, this was only half the problem—when the tungsten filament glowed, its atoms started to evaporate under the heat, making the wire thinner and considerably shortening its life. On top of this it also produced condensation inside the bulb, which in turn reduced the

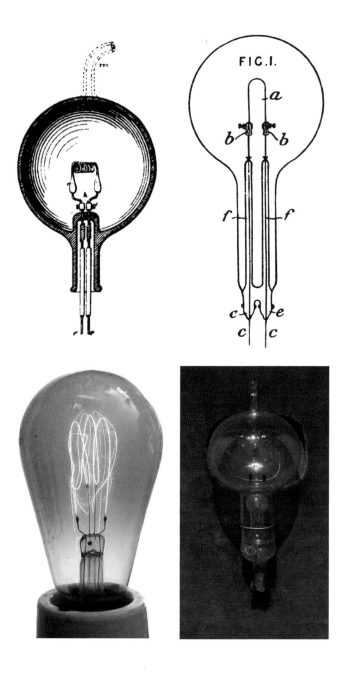

FIG.I.

light emitted from the filament. Another fellow American, Irving Langmuir, came up with a solution to this problem. In 1913 he coiled the tungsten wire filament around a supporting bar, thus reducing the amount of heat omitted and increasing efficiency. He also filled the glass-bulb vacuum with inert gas such as nitrogen and argon, slowing down the process of evaporation and decreasing the amount of condensation. The result was the common bulb that is still used today.

A by-product of the race to create a better bulb saw French chemical engineer George Claude inadvertently create neon light. Claude was disappointed when he passed an electrical current through a tube of neon gas, without the use of a filament, and obtained a glowing red light, but a keen-witted advertising agent could see the potential in such an invention. Jacques Fonseque envisaged that the glass tubes could be bent to form letters and used as coloured signs for shop fronts at night. The first of these to be erected was on the front of a Parisian hairdressing salon and the next for an advertisement for Cinzano in 1912. Other gases were used to give varying colours, including mercury vapour for a

blue-white light; however the light produced actually consisted of harmful ultra violet rays. It was not until the 1930s that scientists discovered that by coating the inside of the tube with a chemical called phosphor, the bulb was able to emit safe light at a quarter of the energy consumption of a normal incandescent bulb. The bulb was exhibited in various colours and sizes at the 1939 New York World's Fair and was an instant hit. By the 1970s the low-voltage, long-life fluorescent light bulbs were being used for the bulk of all business and industry needs.

Ecological concerns regarding energy use in lights have become increasingly prevalent in today's society. A standard incandescent light bulb wastes 95 per cent of the energy used as heat, with only five per cent going towards lighting our homes. It was Ed Hammer who provided an energy-saving alternative through his "compact fluorescent light bulb". A standard light bulb will last just five months, but a CFL will last twelve times that and requires only 20 per cent of the energy that the incandescent bulb uses to provide the same light.

1679
Denis Papin

PRESSURE COOKER

Left Example of a modern stove top pressure cooker.

Top right A commercial steam pressure cooker, 1921.

Bottom right A recognisable version of the modern day kitchen pressure cooker. Submitted by Duane H Walker and Darrell W Quarderer and published on 31 July 1979.

French physicist and mathematician Denis Papin invented the pressure cooker—or the "steam digester" as he called it—in 1679, using the principles of a law determined by Robert Boyle, for whom Papin was working as an assistant. Boyle's Law describes the inverse relationship between the pressure and volume of gases: when a gas is put under enough pressure it will condense into liquid, and if this liquid is heated under enough pressure its boiling point will rise significantly. Papin produced a pot with a lid that could be tightly sealed to allow for the build up of pressure; included was a valve to control the pressure, which opened and closed to release excess steam. The pressure cooker could reach a top boiling temperature of 120 degrees Celsius, cooking at three or four times the rate of a normal pot. It was especially good for cooking tough meat and softening bones. A year after its invention Papin used his steam digester to cook for King Charles II and the Royal Society, who were so impressed they made him a member.

Papin can also be considered the godfather of the steam-powered Industrial Revolution and the steam locomotive. It was through the pressure cooker that Papin realised that steam could be utilised as a power source to pump a piston: the steam produced by boiling liquid pushed the piston up, while cooling the exterior of the pressure cooker with cold water causing the steam to condense, which in turn caused the piston to fall. The repetition of this cycle would continuously pump a piston and power a machine. Papin designed such a machine but it was incredibly impractical and was probably never actually produced.

Modern pressure cookers have, for the most part, stuck to the original design: a lid that has a rubber seal locked onto the pot and a safety valve with an indicator to show and control pressure. Usually there are only two settings to choose from: a low pressure for foods such as fish and vegetables, and a second high pressure setting for everything else. The benefits for using a pressure cooker include its speed, energy saving abilities and the fact that the food does not need to be stirred.

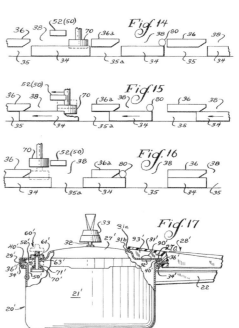

AGA COOKER

The now famous design icon and British favourite, the AGA cooker, is in fact of Swedish origin; designed by engineer and Nobel Prize winner Gustav Dalén in 1922. The story of the development of the AGA cooker is both tragic and romantic: after an unfortunate accident occurred whilst working on an earlier invention, Dalén was left blind and his wife was left with an extremely heavy workload, both having to care for her husband and continue the day-to-day running of the house. Dalén recognised this and set about designing something that would relieve his wife's stress and allow the couple to spend more quality time in each other's company.

The AGA (Aktiebolaget Gas Accumulator) cooker was Dalén's solution. The AGA used a fuel source that burned slowly yet effectively, relying on its heavy iron castings to retain heat. Originally fuelled by coke, today the cooker can use gas, petrol, diesel, electricity and even solid fuels such as coal or wood. The AGA was not only aesthetically pleasing but also extremely practical and versatile. Cooking large meals was comparatively easy with the AGA because it was large and contained two ovens—the top oven being hot and the lower one warm. Similarly, the left-hand hotplate was very hot for boiling and frying, whilst the right-hand hotplate was less hot for simmering. Not only could you cook a variety of foods at the same time, but also because the food was cooked through radiating heat rather than heated air, it retained its moisture and flavour.

The practical uses of the AGA did not stop within the culinary sector; as well as heating the house it could be used for drying laundry on the rails attached to the cooker itself or on a drying rack hoisted above the AGA. When compared to most conventional, modern-day electric or gas ovens, the AGA excelled as a device in a house that was constantly occupied.

Every AGA is made solely in Coalbrookdale, Shropshire, England and they have proved hugely popular in Britain since their introduction 80 years ago. They are often affectionately referred to as the "heart of the home".

1945
Dr Percy LeBaron Spencer

MICROWAVE OVEN

Top right An early concept for the heating of food through the use of electromagnetic energy, submitted by Percy LeBaron Spencer in 1945, though not passed until 1950.

Bottom right A 2004 design by Yoon Gon Kim, assigned to Samsung.

The microwave oven was developed from an incident involving military radar technology during the Second World War and a chocolate bar in a researcher's pocket. Radar technology using microwaves had been developed by the British during the war to detect enemy craft, but even in these early stages of research, scientists in Birmingham had noticed that the microwaves produced by the magnetron device radiated heat—so much so, that they used the beam to light cigarettes off.

It was Dr Percy LeBaron Spencer, a top engineer researching radar-based technology for the Raytheon Corporation of America that realised microwaves' capacity for domestic use in 1945. Working within close vicinity of the high-frequency radio waves produced by magnetrons, he found that a peanut-based chocolate bar he had kept in his pocket had melted. Guessing that the microwaves may have caused this, he sent out for a bag of popcorn and placed the bag next to the magnetron. The kernels inside immediately began to pop. His next experiment involved a kettle, an egg, and the face of a fellow employee. Spencer cut a hole in the kettle and then placed a raw egg (still in its shell) inside. He then situated the magnetron next to the kettle; at this point an intrigued colleague peeked in for a closer look and the hot egg exploded, leaving him—literally—with egg on his face. The process relies on the fact that when the microwaves pass through the food, its molecules absorb the energy (especially water molecules). Each of these molecules has a positive and negative charge. As they try to align themselves with the magnetic field produced by the microwaves, they consequently produce movement and friction, which in turn produces heat. It is essentially this agitation that cooks the food.

Raytheon immediately filed a patent for Spencer's discovery and produced the first working prototypes to be trialled in Boston restaurants in the same year. It took a mere two minutes and 20 seconds to cook a whole chicken, and an astonishing 60 seconds to cook a hamburger. The first model that was put on the market for commercial use was the size of a fridge and weighed 750 pounds; it retailed from $3,000 to $5,000 (around £28,000 to £46,000 in today's money—you can now buy a good microwave for as cheap as £35). More worrying, however, was that they were prone to leaking radiation. In 1965 a smaller model called the "Radarange" was put onto the market for domestic use; it was smaller, air-cooled and cost a 'mere' $500.

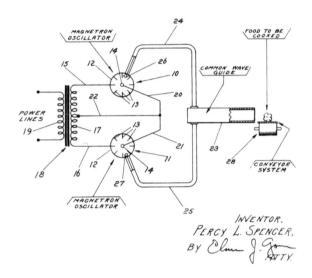

Jan. 24, 1950   P. L. SPENCER   2,495,429
METHOD OF TREATING FOODSTUFFS
Filed Oct. 8, 1945

INVENTOR.
PERCY L. SPENCER,
BY

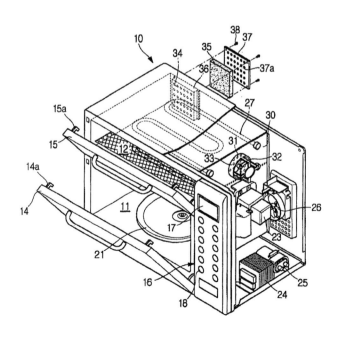

READY MEAL/TV DINNER

The Ready Meal/TV Dinner is a whole meal neatly packaged in a tray, which can be heated up in a matter of minutes for a quick culinary fix. The Ready Meal's popularity accelerated with the help of other technological advances such as the microwave, and served to meet a growing need for a quick solution to preparing a meal in an increasingly hectic world.

The actual identity of the inventor of the first Ready Meal is open to question. In 1996, Gerry Thomas, the retired executive at Swanson—maker of the "Swanson TV Dinner"—laid claim to the idea, implying that a surplus of frozen turkeys after Thanksgiving had prompted him to come up with it. However, this has been hotly disputed by members of the Swanson family, alongside, *The Los Angeles Times* who claim that it was in fact the Swanson brothers themselves who came up with the invention.

But regardless of this, the Ready Meal had already existed in a rudimentary form beforehand. In 1944, William L Maxson's frozen dinners were being served as in-flight meals, and in 1948 frozen fruit and vegetables were sold under the label of "Dinner Plates". In 1952, the first frozen dinners packaged on oven-ready trays were introduced by Quaker States Foods, and were swiftly copied by Frigi-Dinner. However, Swanson were able to use their nationally recognised brand name (they were already established as a large producer of canned meat) to launch a major campaign using the name "TV Dinner", capitalising on the excitement surrounding the television at the time.

Because of the need to ensure that pre-packaged Ready Meals have some semblance of flavour, they are often pumped with salt and fat, alongside partially hydrogenated vegetable oils to preserve the product for a long period. Recently, however, the trend has been in the other direction—particularly in the United Kingdom—where low sugar, salt and fat alternatives are increasingly used in prepared foods.

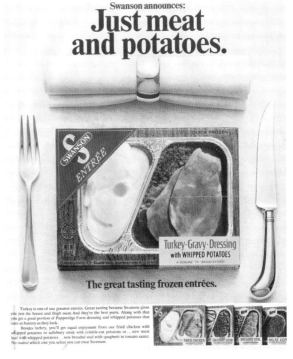

Top An advertisement for Swanson TV Dinners.

Bottom An example of a Hungry Man TV Dinner, produced by Swanson.

POP-UP TOASTER

Top and middle This patent, dated 12 September 2000, depicts the inside 'popping' mechanism of a toaster, including a push-down handle and interior platform positions. An updated mechanical design, showing the toaster's interior mechanisms in greater detail is also shown, dated 13 August 2002.

The discovery of electricity in the mid-1700s saw many inventors trying to find practical ways in which to harness its power. During the Victorian period the potential for the use of electricity in the home—especially the kitchen—was realised. The early twentieth century house became electrified, eradicating the dirt and smoke of previous centuries from the kitchen. Many appliances were developed to take full advantage of electricity as a power source within the home. The pop-up toaster was one such appliance.

The Romans were very fond of what they called *tostum*, which involved scorching or burning the bread by holding it near to an open fire using a fork or tongs. This process did not change dramatically until the advent of electricity. The first toaster was a caged unit that sat on the table face-up and was heated using current-resisting elements from electric fires and irons, designed by British company Crompton & Co in 1893.

As the toaster underwent refinement, the procedure of placing bread facedown on the rack and turning it periodically was the next thing to go. American Lloyd Copeman designed a toaster with sprung doors that turned the toast automatically when they were opened. However, the patent of 1914 was not issued to Lloyd, but his wife Hazel, who asked while window shopping one day: "Lloyd couldn't you invent a toaster that would automatically turn the toast?"

The one remaining problem was that a watchful eye still had to be kept on the toast in order to avoid burning, since the doors had to be sprung manually. It was American Charles Strite's desire to avoid as little manual attention as possible that led to an invention that stood the toast up vertically on a spring mechanism, which was released by a variable timer to make the toast 'pop up' when finished. He filed a patent for the device in 1919 and originally only foresaw its use in restaurants, but by 1926 the 'Toastmaster' was very popular with a huge portion of the American population. The British did not really pick up on the idea until the first Morphy Richards model was produced in 1948.

Worth noting is the importance of another invention without which the pop-up toaster may never have caught on in the way it did. In 1928 American jeweller Otto Frederick Rohwedder invented a machine that sliced bread into equal sizes and then very quickly wrapped and sealed the loaves in order to capture their moist freshness and stop them going stale. These toast-ready loaves were perfect for everyday use in conjunction with the toaster.

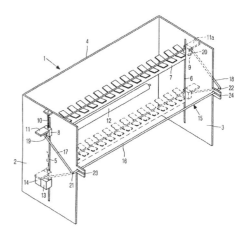

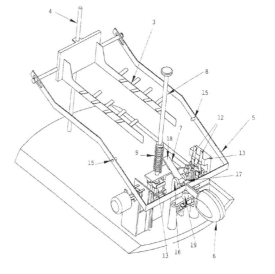

REFRIGERATION

Preserving food with ice to keep it at a cool temperature that is uninhabitable to most bacteria is a practice that goes back to prehistoric times. Collecting snow and ice for this purpose during winter was a common practice amongst most ancient cultures. The Persians dug ice pits for storage, while other cultures used insulated caves; icehouses can even still be found on the grounds of a few stately homes in Britain today.

The discovery of chemical refrigeration in the sixteenth century gave rise to the first method of cooling food artificially. The addition of sodium or potassium nitrate to water caused a lowering of temperature and the resultant cold water operated as a cool bath in which food could be stored. A number of subsequent methods of artificial refrigeration emerged at the hands of Scott William Cullen and American Oliver Evans; but it was not until the experiments in vapour-compression refrigeration by Alexander Twining in 1848, that commercial refrigeration really began to take off.

In the 1870s, breweries were the largest users of commercial refrigeration units, but even then many still relied on ice harvested from natural sources. It was only at the beginning of the twentieth century, when pollution and sewage began to contaminate natural ice that the demand for artificial refrigeration methods began to take hold—particularly in heavily built up areas.

In 1895, German engineer Carl von Linde designed and set up a process for the production of liquid air and subsequently liquid oxygen on a large scale for use in domestic refrigerators, and by 1911 the domestic refrigerator began its ascent to ubiquity. Just three years later, in 1914, commercial refrigeration had surpassed natural ice to become the norm.

TETRA PAK

When Dr Ruben Rausing founded Tetra Pak in 1951 he turned his attention to finding ways in which to package liquids in order to maximise their longevity. This process, known as 'aseptic packaging', resulted in the Tetra Classic in 1953. Its innovative design allowed liquids already subjected to ultra-high temperature processing (UHT) to be packed in a vacuum and then stored for up to one year. The success of Rausing's design was in his approach—his philosophy of minimising packaging while maximising hygiene was reflected in the ethos of the Tetra Pak; the shape of which was designed to 'wrap around' the liquid and give nothing more. The tetrahedron-based form allowed no space for air or oxygen, which would have had a damaging effect on the enclosed liquid.

This idea evolved into the Tetra Brik in 1963, which introduced the familiar rectangular design that is still used in long-life milk cartons today. Tetra Rex then took this further by introducing the gable-topped design, which is still in use and remains equally loathed as it is loved. It relied on the folding mechanism that allowed the top of the packaging to be rearranged into a pouring lip, the mixed success of which often meant that the user could end up pouring liquid inaccurately. Despite its flaws, the efficiency of this folding system remains undeniable and its influence wide reaching.

The innovations in liquid packaging that Tetra Pak introduced have had such widespread influence that, before his death in 1983, Ruben Rausing was Sweden's richest person and his company—once a small family-run affair—had become one of Sweden's largest corporations.

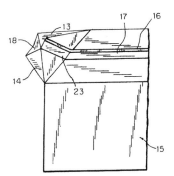

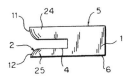

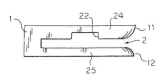

Left A 'modern' American fridge/freezer.

Top right A patent illustration of a closing clip for the a pouring spout style Tetra Pak container, dated 25 March 2003.

WASHING MACHINE

The basic principle of washing clothes or linen hinges on the forcing of water through garments, with the addition of a cleaning substance. Passengers on board ship voyages used to put their clothes in a tough cloth bag, attach the bag to a rope and then throw it overboard and pull it through the sea at speed for a few hours. Unfortunately, back on dry land, the power needed to clean clothes had to be produced by elbow grease alone. The weekly wash for the family could take a housewife all day, and not only was it time consuming; it was also extremely tiring and laborious. Water from a well or valve tap had to be transported to the stove to be heated to high temperatures; a single series of washing, boiling and rinsing could use up to 50 gallons of water and buckets could weigh 40 pounds. These heavy, wet clothes and linen then had to be lifted, turned and kept in constant motion using tongs, finally being wrung and mangled, then hung out to dry.

The first electric washing machines had to replicate the movement of water and heat it to the right temperature, which usually used several kilowatts of power. The first person to produce an electric-powered washing machine capable of this was American engineer Alva Fisher who, in 1906 invented a basic motorised tub that stirred the washing, nicknamed the "Thor". Fellow Americans, Rex Bassett and John W Chamberlain, invented a far superior machine, more in line with the modern-day washing machines; the major difference being that the Bendix 'Model S' of 1937 was automatic, needing little human intervention during the cycle. However, it was very unstable at high spin speeds and needed to be bolted to the floor. In part because of the refinement still to come in washing machine design (such as shock absorbers) and also because of the post-war demand for consumer goods, washing machines did not really become popular until the late 1950s/early 1960s. It was the 'twin tubs' that really took off, which featured one tub for washing and one tub for spin-drying, encased in a single cabinet.

Advancements and refinements in washing machine technology continued, with the machines becoming much more proficient in their task. Dyson brought out their 'Contra-rotator' washing machine in 2000, which had two drums spinning in opposite directions for improved cleaning; each drum also containing 5,000 perforations for enhanced drying capabilities while spinning. Today, washing machines have also had to become increasingly efficient at washing clothes at lower temperatures, as ecological issues become more prominent in today's society.

Right A patent by Alva Fisher for a washing machine drive mechanism, 9 August 1910.

DISHWASHER

By 1900, Josephine Conran, an American housewife from Shelby, Illinois, had successfully mechanised the washing of dirty dishes and achieved her goal of relieving tired housewives from the chore. Furthermore, she had achieved this by inventing a machine whereby her fine china would not be chipped or, worse yet, broken by her servants. On 18 December 1886, Conran received a patent for her mechanical washing machine. It had wire compartments for holding plates, cups and saucers that were then placed in a covered cauldron, and the machine was hand-cranked to spray water over the dishes. Latterly a motor turned the wheel that lay flat within the copper boiler as hot, soapy water was sprayed upon the dishes. Mrs Conran showed her invention at the 1893 Chicago World's Fair, but only restaurants and hotels were really interested, and it was not until after her death that the dishwasher became popular with the average housewife.

Hardware merchants Willard and Forrest Walker made dishwashers with either a petrol or electric motor in 1912, but they were very expensive, leaked profusely, and were not remotely reliable. Unsurprisingly, the dishwasher's popularity and development ground to a halt until the General Electric Company bought the Walkers out, and an effective detergent especially designed for dishwashers was developed—Calgon—in 1932. Shortly after, the first real boxed-in dishwashers appeared, and by the 1950s automatic machines with rinse cycles were on the American market and were introduced to Europe in 1960.

Dishwasher Aids Housewife

With power supplied by a small electric motor, this dishwasher cleans an entire filling of dishes in eight minutes and uses but six quarts of water. Stationary dish holder prevents breakage.

AN ELECTRIC dishwashing machine which uses six quarts of water, cleans all the dishes in the machine in eight minutes. Taking up but little room in kitchen, the mechanism is simple enough to be operated by a child.

The dishes are placed in a basket which in turn is placed in the machine. The basket is self-locking, and stationary during the washing operation, thus eliminating all chance of dish breakage. An agitator with four blades revolves around the perforated basket, forcing the water upward between and over the dishes. The water strikes the dishes at all angles, doing a thorough and sanitary job.

Now more than ever

Give her a Hoover

and you give her
the best _____

"Twas the night before Christmas"
And the last gift of them all—a man's
gift to his wife . . . for a merrier
Christmas and happier years to come.

★

Smart cellophane wrappings hide something thrillingly new in helpfulness. Lucky woman!

Her husband's giving her the Hoover One Fifty Cleaning Ensemble . . . the first basically new cleaner in 10 years. Now she'll clean everything as she goes . . . rug-and-furniture cleaner in one ensemble, instantly convertible. Now she'll clean with new ease . . . a new wonder-metal, magnesium, gives new lightness. Now she'll have a cleaner crisply modern . . . new Henry Dreyfuss design. Now she'll have at her beck and call conveniences never before known to cleaners . . . 15 major improvements. Plus the rug-protecting cleaning action that only the Hoover can give—patented Positive Agitation.

Another fine Hoover for Christmas Giving . . The sensational Hoover "300". A full size, precision-built Hoover, with the exclusive, patented cleaning action, Positive Agitation, the electric Dirt Finder and many other basic $1.00 a week Hoover features, at an amazingly low price . . . *payable monthly*

These and Other Exclusive or Patented Features

Handy Cleaning Kit
Automatic Rug Adjuster
Time-to-Empty Signal
Clip-on Plug
Spring-Cushioned Chassis
Instant Bag Lock
Plug-In Tool Connector
Non-marring finish in stratosphere gray

Your Christmas Hoover will be delivered already wrapped in gay holiday cellophane, if you wish. Sold by leading stores everywhere, through responsible representatives, $1.50 a week and on easy terms *payable monthly*

THE HOOVER *One Fifty* CLEANING ENSEMBLE

IT BEATS . . AS IT SWEEPS . . AS IT CLEANS

Top The Dyson bagless vacuum cleaner, the first vacuum cleaner invented by James Dyson.

Bottom Patent for the Dyson vacuum cleaner, filed 14 April 1979.

VACUUM CLEANER

The first machine to use the principles of a vacuum for cleaning purposes was invented by English civil engineer Hubert Cecil Booth in 1901. During a train journey in America, Booth witnessed a steward futilely trying to clean the upholstery of the seats with a device that blew the dust away using compressed air, only to see it fall back down and settle on the momentarily clean surfaces of the furniture. Subsequently Booth came up with the idea to reverse the force, sucking the dirt and dust from the surface. When he got back home Booth tested his theory by stretching pieces of cloth over his mouth, kneeling down to the floorboards and sucking hard with his lips an inch away from the floor. However ridiculous a spectacle this may have seemed, it worked: the cloth was covered in dust.

Booth began developing his vacuum cleaner and in 1902 sent teams of employees from his British Vacuum Cleaner Company out door-to-door to sell the contraption. This was no mean feat for the teams. Powered by a five-horsepower oil engine, the 'Puffing Billy' was huge and had to be carried by horse and cart. It would be deposited outside the houses of nobles, and an 800 foot hose passed through the windows in order to clean their carpets. Inevitably, it did not really catch on, and neither did the down scaled, electric-powered 'Trolley Vac', still weighing in at a hefty and not particularly portable 40kg, released two years later.

James Murray Spangler of Ohio received a patent for a far more conventional vacuum cleaner in 1908. Spangler was an asthmatic janitor and was desperate to design a machine that meant he could clean without running the risk of an attack. His idea was based on sucking the dirt into a bag rather than through a filter at the front of the machine. The first prototype was made from an old soap box sealed with tape, an electric fan and a pillow case; all connected to a stove pipe attached to a broom handle, which in turn was attached to a paint roller coated in goat bristles. After some considerable development Booth sold the rights to his cousin's husband, William H Hoover. In the same year, the Hoover Company released the Model 'O' vacuum cleaner.

The basic design of vacuum cleaners stayed much the same for some 70 years, aside from refinements in design as technology progressed. It was not until 1978 when James Dyson had an idea that would completely change the way in which vacuum cleaners operated. While cleaning his house Dyson ran into a problem: the bag was full and he had no replacements. Dyson resigned to emptying the old bag and reusing it, but when he opened the bag it was hardly full at all. After cleaning it out and fitting the bag back in place he began again, only to observe that the machine's suction power was still poor. The walls of the bag had become saturated with dirt and were inhibiting the machine's performance. He began designing a bag-free vacuum cleaner that relied on centrifugal force to spin the dust and dirt out of the air, optimising suction. After 5,000 prototypes and 15 years of development the Dyson DC01 was put on to the market in 1993, relying on a "dual-cyclone system that accelerates the dust-laden air to high speed, flinging the dirt out of the air-stream". It was an instant success, and remains one of the most popular machines of recent times.

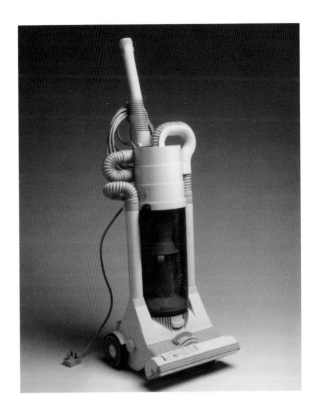

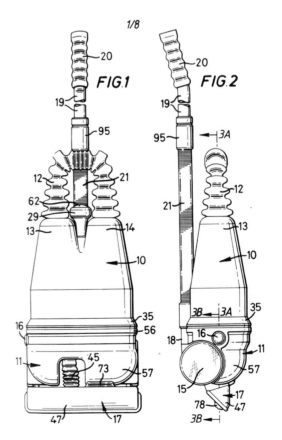

CENTRAL HEATING

The first evidence of a system for diffusing heat has been found in the ancient ruins of the palace of Knossos, in Crete. The resident Minoan civilisation were the same people to have devised a rudimentary flushing toilet. Although the Ancient Greeks may have been the first to use such a system, the Romans were the real engineers of central heating. Due to the cold winters of the north—especially in Germany and Britain—the Romans invented the 'hypocaust' system: tiled floors were suspended upon a series of columns to create a void between the floor and the foundation floor. Into this space hot air was dispersed from a central fire, travelling from one room to another by the use of interconnecting flues in walls. These conduits had to be made out of a quality material—wood and earthenware would not suffice—so they were made out of lead. Ancient central heating also inadvertently created the job of the plumber.

However, with the fall of the Roman Empire, central heating also fell out of fashion. It was not until the Industrial Revolution of the early 1800s that central heating was used again to heat large buildings for industry. It is in this way that central heating is directly linked with air-conditioning: both are systems for interior climate control. They work in conjunction to form HVAC systems (Heating, Ventilating and Air-conditioning), providing specific environments for specific needs, with most modern work environments being controlled in this way. Modern houses are centrally heated. Hot water is pumped from a boiler into a system of radiators, usually controlled by an automatic timer and thermostat. Although this system is the most widely used, the Roman idea for under-floor central heating is considered far better as there are no draughts and no hot and cold spots within the house. Not surprisingly, many private new builds opt for under-floor tube heating rather than wall-mounted radiators.

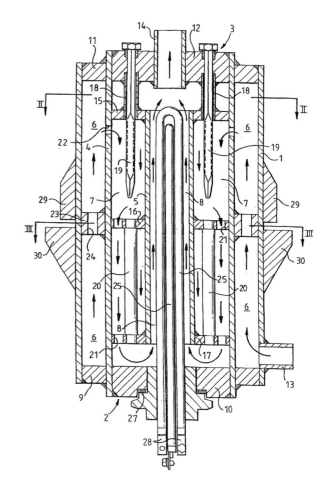

**1906
Willis Haviland Carrier**

AIR-CONDITIONING

Barely a year after graduating from university
with a Masters degree in Engineering in 1901,
New Yorker, Willis Haviland Carrier had gone
from working at the Buffalo Forge Company for
$10 a week, to becoming head of their department
of experimental engineering and inventing the
first air-conditioning system. He was hired by a
Brooklyn publishing firm in 1902 to design a system that would
stabilise air conditions within their printing plant; the weather
outside was causing fluctuations in temperature and humidity,
which in turn was causing the dimensions of the paper to alter
slightly and misalign the printed colour inks. Carrier's "Apparatus
for Treating Air" pumped liquid ammonia through a set of
evaporation coils; the ammonia then absorbed the heat of the
warm air in the room and evaporated. The air was finally cooled
until it reached its 'dew' point and the droplets drained away,
leaving the cooler and drier air to return to the room via a fan.

The invention was granted a US patent in 1906 and in 1911
Carrier disclosed his basic "Rational Psychometric Formulae" to
the American Society of Mechanical Engineers, allowing anyone
to use it. This led to the proliferation of air-conditioning units
manufactured for all sorts of industrial uses. The ability to control
interior conditions of heat and humidity improved the quality of
numerous products including tobacco, food and film. The system
was refined for commercial use in 1924 and Carrier's "Centrifugal
Chillers" were installed in department stores, cinemas and theatres.
The venues saw a boom in trade, as did the Carrier Company,
which experienced a huge demand for these smaller units. One
hundred years after its conception, air-conditioning can be found
in planes, cars and almost any place of business.

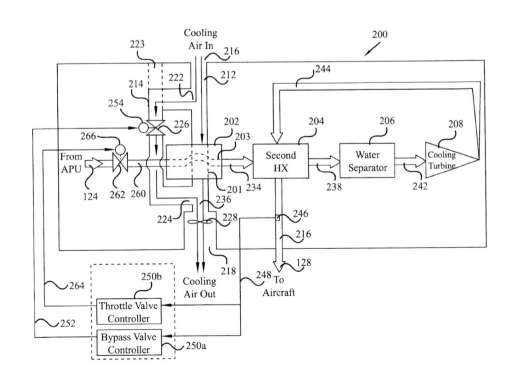

SOAP

Soap supposedly originated from a substance of animal fat and wood ash that was washed down from the Roman sacrificial site of Mount Sapo and deposited on the clay banks of the River Tiber. However, clay cylinders have also been found that had a soapy residue in them dating back to 2800 BC—it has hence been thought that the boiling of fat and ashes to produce a cleaning substance was a frequent Babylonian and Egyptian practice as well.

The method of making soap lies within a basic chemical reaction that occurs in the neutral fat of a strong alkali—one end of a soap molecule is hydrophilic (attracted to water) and the other is hydrophobic (repelled by water), but is attracted to oil and grease. When the surface of the soap becomes wet, its molecules attract oil and grease and hold them in suspension until rinsed away.

By the 1500's vegetable oils began to be used instead of animal fats, with caustic soda gradually replacing wood ash. Bars of soap began to be produced in the late eighteenth century as advertising campaigns began to encourage the relationship between cleanliness and health in the industrial cities of the world. In London, for instance, Andrew Pears began making his world renowned, transparent soap in 1789. The chemistry of soap making remained much the same until, in 1916, the first synthetic detergent made from a variety of raw materials was produced in Germany in response to the lack of fats available during the war.

THE ORDER OF THE BATH

Ancient Greece

SHOWER

The showers of Ancient Egypt involved servants standing behind a low wall, pouring water over the washer. The Ancient Greeks, however, invented the first showers with plumbed-in water. A scene glazed on an Athenian vase dating from the fourth century BC depicts four women showering in two different shower rooms; piped-in water is sprayed upon them through perforated shower heads located in the mouths of mounted ceramic heads of boars and lions. These shower rooms were most probably used by the Greek athletes of the Olympics as a fast way of washing and freshening up. In Pergamum, a Greek metropolis in western Turkey, a complex of seven showers was unearthed that dates back to the early second century. The system was supplied with water through an overhead main, which was then sprayed on to the bathers. The water then drained into a foot well which, on reaching its maximum capacity, flowed into the next shower unit and so on, until it reached a drain. The footbath is such a successful feature that it is found in almost all modern shower units.

**1597
John Harrington**

THE FLUSH TOILET

The necessity of urban sanitation systems increases with the rise in population of a burgeoning society that inhabits a certain town, city or region. Without sanitation, the risk of disease increases significantly. Remnants of such systems and units have reportedly been found by archaeologists in the ruins of the cities of Ancient Crete dating from around 2000 BC. Flushing technology including a system of clay pipes and sewers, rudimentary manhole covers and a set of stone closets have been excavated at the Palace of Knossos site.

The first attempts at sanitation in European society, however, came during the late Tudor period of England's history when in 1597 John Harrington, Godson of Queen Elizabeth I, invented his own flush toilet that he nicknamed the "Ajax". It consisted of a cistern that released water into the toilet bowl, washing away the excrement through a trap that was emptied into a repository. The Queen had one installed in Richmond, but the idea did not catch on. Over the following two centuries the populations of England and Europe continued using their 'privies'—a basic bucket in a closet that was thrown out into the street after use. The common name for a toilet, "loo", is derived from this unhygienic practice. The French used to call out *garde a l'eau!* before tossing out their waste, warning people to "watch out for water".

One of the most widespread myths about the flush toilet is that a Mr Thomas Crapper invented it. Crapper was an exceptional plumbing engineer but in fact manufactured the 'S' bend. He was so accomplished at this job that it earned him a royal warrant for providing toilets for the monarchy. The first legitimate patent for a modern flush toilet was awarded to Alexander Cumming in 1775, a watchmaker by trade who invented a toilet with an 'S' bend. Its ingenuity lies in its simple conception: the S-shaped, curved tube prevented backflow of waste and unwanted smell, providing an air and water seal between the bowl and sewer. However, there was still no real market for the flush toilet until the 1800s when the potteries and iron foundries began to improve terracotta and iron pipe respectively. More influential was the instatement of The Public Health Act of 1848, which saw five million pounds invested into sanitary research and engineering in Britain. Increasingly, flush toilets were made from ceramics, often beautifully decorated by artists at manufacturers such as Armitage Shanks and Thomas Twyford (who created the first one-piece ceramic toilet during the 1870s). By the end of the nineteenth century the new English toilet designs had spread across the Atlantic and gained popularity among its citizens, setting the ball rolling for their instalment and use across the world.

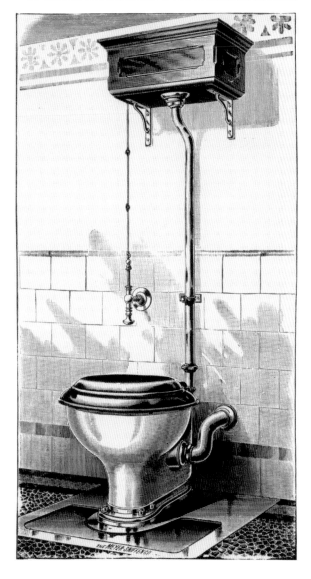

TOILET PAPER

When China's Bureau of Imperial Supplies first began producing toilet paper in 1391 for the use of the Emperor Zhu Yuan Zhang during the Ming dynasty, each individual sheet was an impressively large three by two feet. As the centuries passed, the size of toilet paper was unsurprisingly scaled down in size. The Gayetty Firm made the first factory-made, mass-produced toilet paper in 1857. New Yorker James Gayetty's 'therapeutic paper' contained high levels of aloe for comfort and each pack contained approximately 500 sheets, each sheet having the company's name printed upon it. It retailed at 50 cents.

Fellow New Yorker Zeth Wheeler patented rolled and perforated wrapping paper in 1871, which would eventually contribute to the further improvement of the practical design of toilet paper. He sold his product through his Rolled Wrapping Paper Company in 1874, but this venture proved entirely unprofitable and Wheeler had to re-think and reorganise his company. He began selling 'The Standard' toilet paper on a roll in plain brown paper packets through the newly renamed Albany Perforated Wrapping Paper Company in 1877. The design spread around the world and is still the basic model for the manufacture of all modern toilet paper. Today, there are many variations in quality, from coarse, rough, cheaper varieties to luxury paper that is smooth, thick, durable and sometimes even perfumed and embossed.

TOOTHPASTE

TOOTHBRUSH

Civilisations going back thousands of years have used some form of paste or liquid to clean their teeth. The Egyptians apparently used a mixture of the powdered ashes of ox hooves, burnt eggshells, myrrh, pumice and water. The Romans developed toothpaste containing crushed animal bones and oyster shell, and added powdered charcoal and bark to help with bad breath; they also added one prized ingredient to make toothpaste: human urine, which effectively acted as bleach because of its ammonia content. Ammonia is still used in modern-day toothpastes but as synthesised ammonia compound.

Toothpaste became more widely used in Britain during the nineteenth century. In 1824, a dentist called Peabody added soap to toothpaste, and during the 1850s chalk was added. By 1873 toothpaste was being mass-produced by Colgate. It was contained first in ceramic pots and, later, jars which people would dip their toothbrush into in order to retrieve the paste. This unhygienic practice did not go unnoticed by Dr Washington Sheffield of Connecticut; who was the first person to put toothpaste into a collapsible tube in 1892. It was called "Dr Sheffield's Crème Dentifrice".

After the Second World War, synthetic detergents such as sodium compounds and fluoride, which proved to dramatically reduce the development of cavities, replaced the soap in toothpaste. In the 1980s, fluoride toothpastes had soluble calcium fluoride added to them, which provided the basis for most modern toothpastes.

Early forms of toothbrush have been in existence since 3000 BC, including 'chew sticks'—twigs with one end frayed for brushing teeth, animal bones and porcupine quills. The first bristle toothbrush, resembling modern-day toothbrush design, was documented in a Chinese encyclopaedia dating from 1498. The design placed the bristles at a right angle away from the handle; the coarse back hairs of a hog were plucked in order to make the bristles, which were then set into either bone or bamboo.

Eventually the idea became popular throughout Europe and in the late eighteenth century the first mass-production of the toothbrush was initiated by Englishman William Addis. Around 1770 Addis found himself in Newgate Prison having incited a riot. While he served out his sentence, he tried to think of some kind of entrepreneurial endeavour that might make him money on his release. After washing his face one morning he began to clean his teeth by rubbing them with a rag, a common and accepted practice, when it occurred to him how ineffective the method was. He began to focus on ways in which to improve dental hygiene, and fashioned a rudimentary toothbrush from the limited materials available to him. A small animal bone from his evening meal was used as the handle, into which he drilled small holes and added bristles, procured from the prison guard, which he tied into tufts and glued into place. Around 1880, after his release, William Addis began to mass-manufacture his toothbrush.

The synthetic fibre Nylon was used to replace natural, animal hair bristles in 1938 by American chemical company DuPont De Nemours; it was called "Doctor West's Miracle Toothbrush" and carried the slogan "vitally important to victory". The American population had been slow to incorporate daily brushing into their lives; however, during the Second World War strict orders were put in place for the maintenance of good dental hygiene of all soldiers.

Since then we have seen the development of various models of toothbrush all claiming to be supreme in the field of dental hygiene. Battery-powered toothbrushes have gained popularity, with the heads spinning in a circular motion, protecting your gums from vigorous vertical brushing. The most hi-tech of recent designs is the toothbrush that utilises sonic waves and changes colour when the user is brushing too hard.

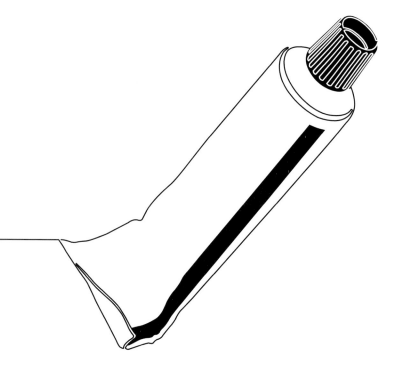

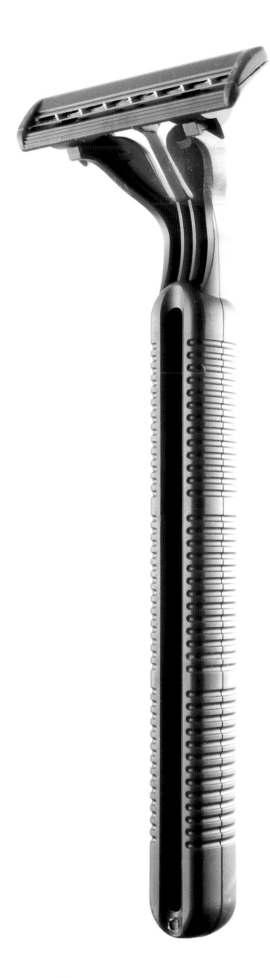

SAFETY RAZOR

Prior to the introduction of the first practical safety razor by King Camp Gillette in 1901, the instrument was more commonly known as the 'cut-throat' razor. Made from tempered steel, its blade resembled a sword; it required constant maintenance and sharpening, and would often cut the user. The design required refining, maintaining a clinical sharpness but introducing a greater element of safety.

Injuries caused by shaving were marginally reduced in 1880, when the American brothers Frederick and Richard Kampfe invented a wire-grid blade cover; the Star safety razor's grid allowed only the facial hair that protruded over the guard to be shaved, though it was still made from steel and as expensive as previous blades.

In 1895, King Camp Gillette, advised by disposable bottle cap entrepreneur and former employee William Painter to "invent something that people only use once, and then throw away" came up with the idea of a shaving razor with replaceable blades. Disregarding professional opinions that a suitable blade could not be made thinly, strongly, or cheaply enough for the concept, Gillette—along with MIT graduate William Nickerson—spent years developing the disposable blade, patenting their design in 1903. Though they became blunt after a few uses, the razors were inexpensive and far more appealing than the high maintenance cut-throat. Gillette's blade almost had the element of safety; only the very edge of the razor protruded from the handle, meaning a drastic reduction in facial cuts.

Modern razors, many still manufactured by the Gillette company, are safer than ever. Designs also incorporate skin exfoliation, by the use of multiple blades eradicating dead layers of skin, and moisturising strips.

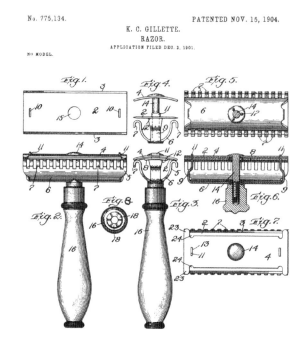

Ancient Egypt

SCISSORS

Early examples of rudimentary scissors have been found in Ancient Egyptian ruins dating back to around 1500 BC. These early incarnations were made out of a piece of U-shaped bronze, the ends of which were sharpened. The C-shape at the handle end acted as a spring for opening and closing. The first scissors that resembled modern-day articles, with cross-blades and a central fulcrum, were developed in Ancient Rome in roughly 100 AD. Commonly used by barbers and clothes makers, these scissors were made out of iron, which was less brittle than their bronze predecessors. During the Roman period scissor-makers were held in very high esteem—they even had their own craft guild.

A major development occurred in 1761 when Robert Hinchliffe of Sheffield gave the cutting instrument durability and efficiency by using cast steel to make scissors. Setting himself up as a high-quality manufacturer in London, he produced beautiful ornate scissors during the 1800s. Most scissors are still made out of steel but in varying degrees: specialised, high-quality variants such as surgical scissors are made out of stainless steel; while cheaper, mass-produced types are made out of softer steel. There are many different types of scissor available today, from 'pinking shears' used by dressmakers and tailors that cut in a zig-zag line in order to stop the fabric fraying, to safety-scissors for children.

Top to bottom Pinking shears; scissors from second century Asia Minor; children's safety scissors.

Ancient Egypt

PERFUME

The Egyptians were the first to incorporate perfume into their culture. From religious ceremonies to embalming the dead, perfume was an important part of Egyptian life. Since the raw materials were so rare—the resins and balms from trees such as Frankincense and Myrrh—it was initially reserved for burning as a sacrificial offering to the Gods; the smoke was believed to create a physical pathway from earth to the heavens. This is how perfume gained its name, from the Latin *per fumum*, which literally translates as "through smoke". The embalming of the dead was seen as an extremely important as it was thought to guarantee safe passage to the heavens. Although at first perfume was used only to worship Gods and anoint Pharaohs, scents created from cinnamon and honey were increasingly used for anointing the bodies of the powerful and wealthy.

The scent of perfume spread throughout the world, and was adopted by the ancient Chinese, Hindus, Israelites, Carthaginians, Arabs, Greeks and Romans. However, it was the discoveries of alcohol, solvent extraction and synthetic chemistry that facilitated the boom in the perfume industry during the nineteenth century. Although perfume was extremely popular in the preceding centuries—particularly in France and Britain—the quantity that could be produced was limited, as was the variety of fragrances. In 1832 J Mero et Boyveau formed a specialist company in Grasse to distil oils and became the first company ever to use solvent extraction (a process invented by Joseph Robert, Henri Robert— who created Chanel No 19, and Guy Robert—who created Caleche and Madame Rochas). This process allowed oils to be extracted in a stable yet intense form from various raw materials, many that had not been available before. More recently, cheaper synthetic ingredients have been used in the mass-manufacture of perfumes, with a consequent cheapening of the end products.

**1926
Erik Rotheim**

AEROSOL

The idea for an aerosol can goes back to eighteenth century Paris and the carbonated soda water that people would buy that was driven out of its containers by carbon dioxide. In 1926 Norwegian Erik Rotheim was the first to establish a product that could be sprayed from a container under gas pressure. The name "aerosol can" indicated that the liquid inside the can would come out in 'aerosol' form (a fine mist). However, until the development of chlorofluorocarbons (CFCs) by DuPont in the 1930s, there was no such gas available that could have been used; furthermore, even after the development of CFCs they could not be dispensed properly without the invention of the disposable can that could contain an aerosol—created by Julian Kahn of New York in August 1939.

The aerosol's first commercial use was during the Second World War when, in 1941, a team working for the US government designed an insecticide spray that utilised the aerosol can. The 'bug bomb' as it was known saved innumerable soldiers lives fighting in the South Pacific by controlling the spread of malaria. Soon after the war—in 1950—Robert Abplanalp designed a plastic push valve that did not clog. His Precision Valve Corporation based in New York made 15 million of them, and it was this that propelled the cheap aerosol can into mass use.

By 1974, the CFCs utilised in aerosol cans came under scrutiny for their damaging effect on the ozone later, and they were subsequently phased out through international agreements a decade later, replaced with flammable hydrocarbons such as butane and propane.

Right Patent illustration dated 29 September 2009, for a 'thermal-type drop generator' that ejects the liquid into small droplets due to its geometry. The diagram shows the expulsion system of the aerosol can and the subsequent droplets of expelled material.

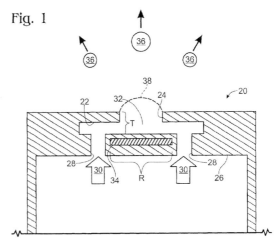

Fig. 1

SEWING MACHINE

Before the development and invention of the sewing machine during the Industrial Revolution, families had to make their own clothes—a laborious process that involved sewing by hand using needle and thread. As early as the mid-eighteenth century people were trying to speed up the process. In 1775, the German, Charles Fredrick Wiesenthal was granted a British patent for his double-pointed needle with the eye situated in the middle, which meant the needle no longer had to be turned around in order to complete the next stitch. Around 100 years later the Wiesenthal needle would be used as a lynchpin component of the first mechanical sewing machine.

In 1790, a patent for a mechanical sewing machine was submitted by London cabinetmaker Thomas Saint. Although it appears from his drawings that the machine would have worked, there is no evidence to suggest that a single prototype was ever built. Over the next 50 years many attempts were made to build a fully operational sewing machine and all failed, apart from one working 'chain stitch' machine, built in 1830 by French tailor Barthélemy Thimonnier. He put his machine into industrial use in 1841 to make uniforms for the French army, but other tailors were less than happy about it. Fearing the invention would put them out of business they rioted and destroyed most of the machines, burning his garment factory to the floor.

It was a poor New England farmer who produced the first automatic lockstitch sewing machine in 1845. After five years of toil and development Elias Howe perfected a system using the double-pointed Wiesenthal needle in conjunction with a double-pointed shuttle, which allowed for a very strong stitch. The machine had no success in America—primarily because the machine cost $300 while the average wage for a seamstress was only $7 a week. Howe travelled to England and sold the British manufacturing rights to corset maker William Thomas, but on his return to America he witnessed the proliferation of sewing machine manufacture. The models were faster, more reliable and cheaper and they were all based upon his own patented system. Howe proceeded to sue these companies (one of them being a company owned by Isaac Merritt Singer, who went on to develop the most popular domestic sewing machine) and won royalties that saw him earn $5 from each unit sold. By the time of his death in 1867 he had earned close to two million dollars.

By the dawn of the twentieth century, sewing machines were being used in households all over America and garments were being made using mass-production lines in factories. Today, little human intervention is needed at all with the invention of computer programmed sewing machines.

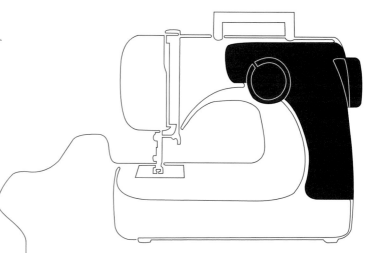

Treadle sewing machine, circa 1920.

1844
Gustaf Erik Pasch

SAFETY MATCH

Invented in China in AD 577 by court ladies of the Northern Qi Dynasty to start fires for cooking and heating, small sticks of pinewood imbedded with sulphur were the forerunners of the modern day match. Further European antecedents date to around 1530, although the first self-igniting match is not formally credited with invention until as late as 1805.

In 1805 Jean Chancel, an assistant chemistry researcher at France's eminent École Polytechnique, created thin wooden splints dipped in potassium chlorate, sugar and gum, which were ignited by dipping into a small asbestos bottle containing sulphuric acid. This match, however, was dangerous and expensive to produce, and it disappeared into near obsolescence without ever gaining much popularity. English chemist John Walker invented the very first friction match in 1827. Walker's match-head mix of antimony sulphide, potassium chlorate, gum and starch was successful in that it could be ignited against almost any surface. The idea was hijacked and patented by Samuel Jones, and the resultant matches emerged with relative success as Lucifer's (light-bearers). Although a breakthrough concept in portable conveniences, the matches suffered from a number of shortcomings—including an unpleasant odour, an unsteady flame, and an unnervingly violent ignition.

In an attempt to remove the match's offensive odour, Frenchman Charles Sauria added white or yellow phosphorus to the match head formula in 1831. His innovation grew in popularity and was rapidly imitated, although the lethal potential of these matches was late to be discovered. Many people involved in their production were subject to physical abnormalities due to exposure to phosphorus fumes, particularly bone deformities—the most prevalent was Phossy Jaw, eventually resulting in the rotting and deterioration of the jaw. It later emerged that a single matchbox contained enough white phosphorus to kill an adult.

Safety matches as we know them today finally came into existence in 1844, invented by Gustaf Erik Pasch in Sweden and improved several years later by Johan Edvard Lundström in 1855. A unique tactic of separating the combustible elements of the match head in combination with a match body suffused with paraffin and a particular striking surface ensured the new match inimitable status. By replacing the original white phosphorus with red phosphorus, Lundström eradicated its fatal potential. Today, safety matches are classified as Dangerous Goods but continue to be available in tobacconists, supermarkets, and airports.

UMBRELLA

Throughout history art has depicted the elite of various ancient civilisations being shaded by a servant bearing a parasol: Egyptian Pharaohs, Indian Buddhas, Greek Emperors, Kings and Queens. The 'parasol' became the 'umbrella' as the idea reached Ancient Rome, at which point it became known as an *umbraculum—umbra* meaning shade or shadow in Latin. The Roman version was neither collapsible nor waterproof, but had started to take on the appearance of a modern-day umbrella—it was made from a wooden frame over which cloth was stretched.

It was in China during the Wei Dynasty of the fourth and fifth century AD that umbrellas were designed that were both collapsible and water repellent. Initially, the use of oiled mulberry paper kept water at bay but, by the fourteenth century, even silk was used—though only the royals were shaded by such luxury. By this time umbrellas were also being used in the Indian subcontinent. In 1340, papal envoy John of Maraignolli spotted one of these umbrellas and brought it back to Florence with him, describing the contraption in a letter to the pope as a "little tent roof on a cane handle, which they open out at will as protection from sun or rain".

The umbrella spread throughout the Western world during the late sixteenth century, and became especially popular in the frequently rainy north of Europe. These early European models were often made out of wood, cane or whalebone with an oiled canvas covering, and were used exclusively amongst women for whom the umbrella had evolved into a beautiful fashion accessory. A man seen using an umbrella faced an onslaught of ridicule. However, Persian traveller and writer Jonas Hanway was not dissuaded by the prospect of mockery, and popularised the use of umbrellas amongst the male population of London in the mid-eighteenth century. For 30 years, until his death in 1786, Jonas carried an umbrella publicly, leading the gentlemen of London to refer to their umbrella as the "Hanway". It was Samuel Fox who designed the metal frame of the modern day umbrella in 1852.

Fox founded the 'English Steels Company', which produced farthingale stays for women's corsets. Fox's reason for designing the steel-ribbed umbrella, replacing the bone or cane model, was to use up surplus stock of these steel stays.

The first shop selling nothing but umbrellas was James Smith & Sons of London on New Oxford Street, which began retailing in 1830 and continues to sell high quality umbrellas to this day. A modern umbrella can be made from varying materials according to taste and cost, but most good quality canopies are made from nylon taffeta; the handle and shaft can be made from materials such as plastic, metal and wood, to varying degrees of quality.

James Smith & Sons Umbrella Shop, based on New Oxford Street, London, is a family run business that has been selling umbrellas to the public since 1830.

THE ZIP

Whitcomb Judson, a mechanical engineer from Chicago patented a 'slide-fastener' or 'clasp locker' in 1893 and exhibited it at Chicago's World Fair in the same year. Whitcomb Judson's device was rudimentary in terms of design, comprising of two rows of large metal teeth that often jammed, came undone, or simply broke after little use. Initially the 'clasp-locker' was marketed for use on high boots, aiming to make the process of putting on and removing the footwear an easier task. In actual fact it made it far more difficult because of the poor quality of the 'clasp fastener'. Inevitably, Judson's invention was not a commercial success, only selling a few units to the US mail as letter bag fasteners.

The invention of the modern-day zip is accredited to Gideon Sundback, a Swedish electrical engineer and scientist, and an employee of the Universal Fastener Company owned by Judson. Sundback was working as head designer at the company plant managed by his father-in-law, when in 1911 his wife's death led him to retreat into his work. From his tragic loss, the world gained the 'separable fastener' in 1913. The trebling of the amount of teeth used on the zip from approximately three per inch to 12 per inch, led to a major improvement in design, action and reliability.

Sundback's separable fastener was used, largely, on the uniforms of soldiers of the US Army during the First World War. The practical capabilities of the invention were not recognised by the fashion industry until the 1930s when clothes designer Schiaparelli first incorporated them into his catwalk show after realising that his models could get in and out of clothes much quicker between changes with the use of the zip. It was around this time that the invention gained its now popularised moniker "zipper" or "zip-fastener" (because of the sound it made in motion) and was gradually incorporated into commercial fashion retail.

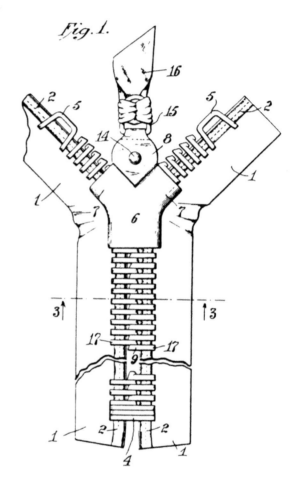

G. SUNDBACK.
SEPARABLE FASTENER.
APPLICATION FILED AUG. 27, 1914.

1,219,881.

Patented Mar. 20, 1917.

Fig. 1.

Fig. 9.

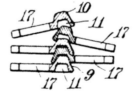

Attest:

Fig. 2.

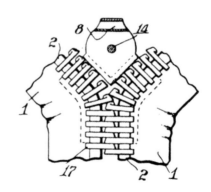

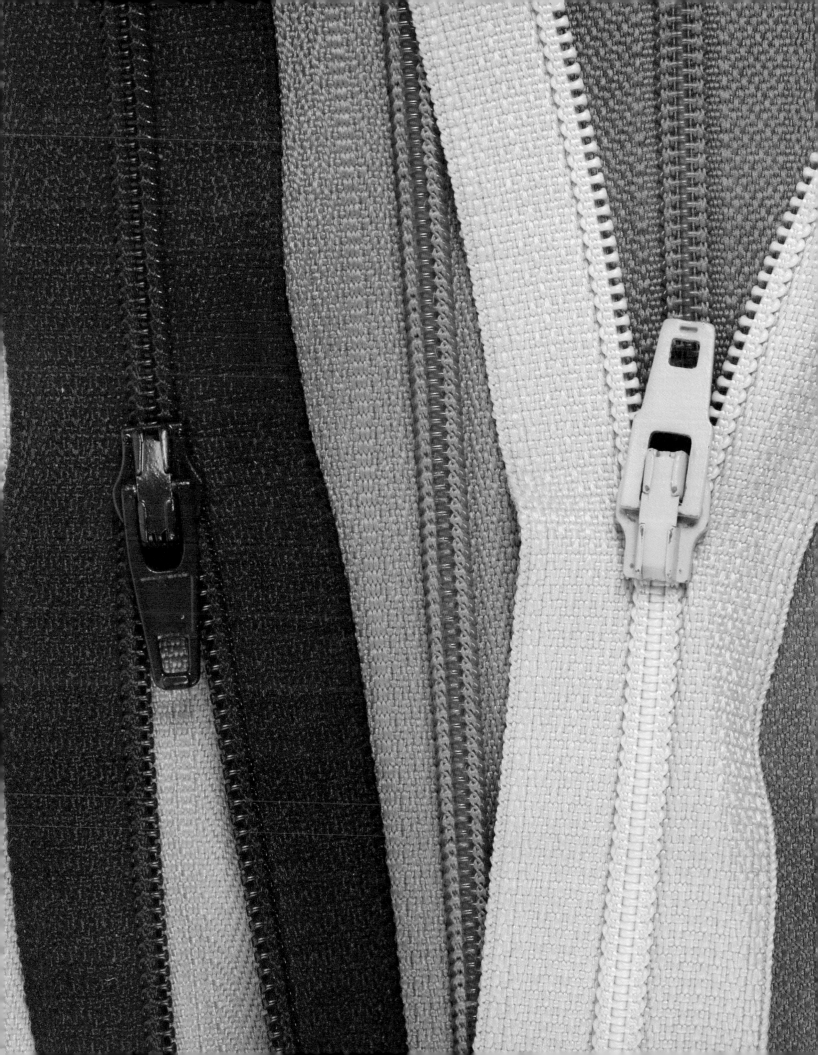

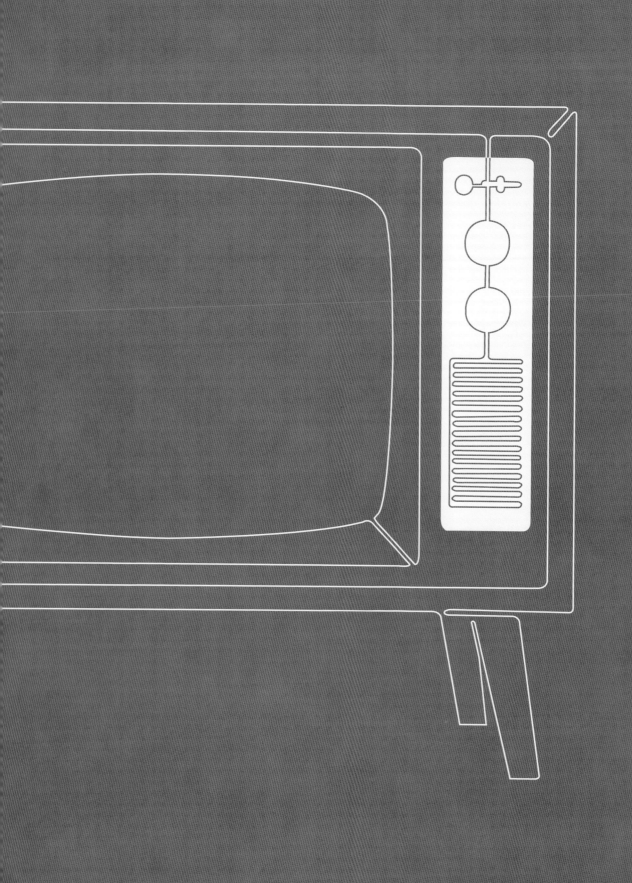

entertainment

PHOTOGRAPHY

Compared with other similar technological developments, photographic techniques were discovered only fairly recently. Literally meaning to "draw with light", photography first surfaced as the camera obscura—a wooden box with a hole in one end that allowed an inverted image of the outside scene to be projected on to the opposite side. It was originally used in the seventeenth century as an artist's sketching aid. The following century saw scientific experimentation with chemical compounds or 'salts'—such as Silver Nitrate—lead to the discovery by German doctor Johann Schulze, that certain compounds darken when exposed to sunlight. Frenchman Louis Jacques Mandé Daguerre unveiled the first practical photographic process at a meeting of the Académie des Sciences in 1839. This involved silver-plated sheets of copper that were made light sensitive by iodine vapour and then developed through exposure to mercury; although still a lengthy process, it was a much-improved version of the eight-hour exposure technique developed by his contemporary Joseph Niépce in 1816.

Around the same time British amateur scholar William Henry Fox Talbot, frustrated by his inability to draw, was developing his own techniques in an attempt to capture the images he saw in his camera obscura. Starting with basic, negative silhouette images of leaves and lace, produced with sensitive paper exposed to sunlight, he managed to fix images with a strong salt solution. It was only a matter of time before he learnt how to use another sheet of sensitised paper in contact with the negative image to expose a positive photograph with the correct representation of light and shade. Frenchman Louis Daguerre also developed a technique of producing bright, clear images, which developed faster than previous attempts—known as 'daguerreotypes'. However, Talbot's method still had a definite advantage over Daguerre's: whilst the Frenchman's technique produced unique one-off images, the Brit's could be used again and again to produce an unlimited number of prints.

Although Talbot's method was undoubtedly preferable, it was still far from perfect: the procedure still involved exposure times of up to an hour and it was not until 1840 that a breakthrough was made: further chemical treatment allowed brief exposures and thus opened up the possibility of portraiture. Following Frederick Scott Archer's discovery of the wet collodion process in 1851, commercial photography boomed, and soon there were photographic studios all over Britain. It was not until the 1880s however that photography reached the masses. In 1888, George Eastman's Kodak revolutionised photography—not only did cameras no longer need tripods and could be held by hand, but exposure times were reduced to a fraction of a second and could be loaded with enough film to take up to 100 photos. Once used, these commercial Kodaks would be returned to the factory, where the film would be processed and the camera loaded with fresh film, before being returned to the owner complete with the set of prints. For the first time, it was possible to take a photograph without any technical skill or chemical knowledge.

Clockwise from top left Number 2 Folding Autographic Brownie; 1951 patent for "photographic apparatus"; Polaroid 1000 Land Camera.

TELEVISION

Radiovision, telephonoscope, lustreer, optiphone, mirascope, telectroscopy—these were just a few of the prospective names circulating during the early development of what we now call "television". The editor of *The Manchester Guardian* at the time declared: "Television? The word is half Greek and half Latin. No good will come of it." Today, however, television has become part of everyday life and can be considered as a central linchpin in the creation of the global village.

By the 1920s the component parts needed to build the first television set had been invented by numerous individuals, and the race was on to see who would succeed first. Many inventors around the world were working in well-equipped laboratories, including Scotsman John Logie Baird and American Charles Jenkins. After being involved in a string of inventions where Baird either narrowly missed out on patent rights, including an industrial diamond-manufacturing process and an air-soled shoe; or inventions that never caught on, like homemade haemorrhoid cream and a glass razor that did not rust; Baird went about designing a TV.

Baird worked by himself, in less than desirable premises, using scrap materials to build a television that worked upon the premise of mechanical scanning. The crux of the design involved a large spinning disc with holes around its circumference—originally patented by Paul Nipkow of Germany in 1884—through which the light passed into a photocell, and the image was scanned and broken down into horizontal lines. The image was then reconstructed in the receiver using the incoming signal to change the brightness of the lamp, which shone through a second scanning disc onto a screen. The image would appear dot by dot, blending into a complete picture. However, to increase sensitivity, Baird made a larger disc—around three metres in diameter—and inserted 20 cm glass lenses into the holes. As befits the slightly peculiar nature of the man, the heavy lenses—spinning at 750 rpm—flew out of their housings and smashed into shards against the walls of Baird's lab. After refining his invention and eradicating the chance of fatal injury, Baird demonstrated his television to shoppers in Selfridges department store in London, who witnessed "a recognisable, if rather blurred, image of simple forms such as letters printed in white on a black card".

DER

6 7 8AV

000

In 1908 Allen Campbell-Swinton envisaged television working on the same premise as mechanical scanning, but using electricity and Braun's cathode ray tube to act as both transmitter and receiver, with a fragmented selium plate as a screen. Each cell of the plate accumulated an electric charge in proportion to the light falling on it, then electronically scanned and created an electric current that varied in time and brightness, producing a moving image on the screen. Eventually, however, a 14 year old farm boy from Idaho, USA, conceived the basic requirements that would herald the invention of the first electronic television that Campbell-Swinton had foreseen. After six years of development, Philo T Farnsworth publicly demonstrated his electronic television system, with no moving parts—the 'image dissector'—in 1929.

Farnsworth was then overtaken and faded from the scene as other inventors made progress during the 1930s. Vladimir Zworykin, whilst working at RCA, developed an electronic receiver tube called the kinescope, and an electronic camera called the iconoscope; with the development of these two component parts the foundations for modern television were set. It then fell to a team of inventors at EMI in Middlesex, England, led by the eminent Isaac Schoenberg to build the first practical television based on the iconoscope: the 'Emitron'.

The first regular television broadcasts began with the instatement of the British Broadcasting Company in 1936, but soon stopped with the start of the Second World War. However, great technological advancements during the war saw television manufacturers embrace electronics and radar to turn out cheap, reliable television sets for the post-war world. By the 1950s the popularity of television was soaring and had become a threat to the world of cinema. Constantly trying to outdo each other, they offered new and exciting developments to enhance viewing pleasure. Since then colour television has been perfected and many technological advances have combined to improve and transform the way in which television is transmitted and viewed—including satellite communications, transistors, microwave relays, screens containing liquid crystals and LCD screens. In the modern era screens are forever increasing in size and are supported by sound systems that aim to create a 'home cinema'. TVs are now well on the way to incorporating high definition, known as HDTV, which uses digital data processing rather than analogue to allow for the compression of images, removing redundant information and providing more space for other programmes and data streams. HDTV contains up to five times the information of a normal TV, which allows for a nearly crystal clear picture.

Clockwise from bottom right Advertising material for a Baird television kit; 1926 Baird television transmitter unit; 1928 model of a John Logie Baird television set.

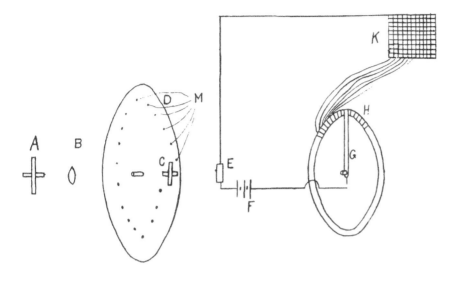

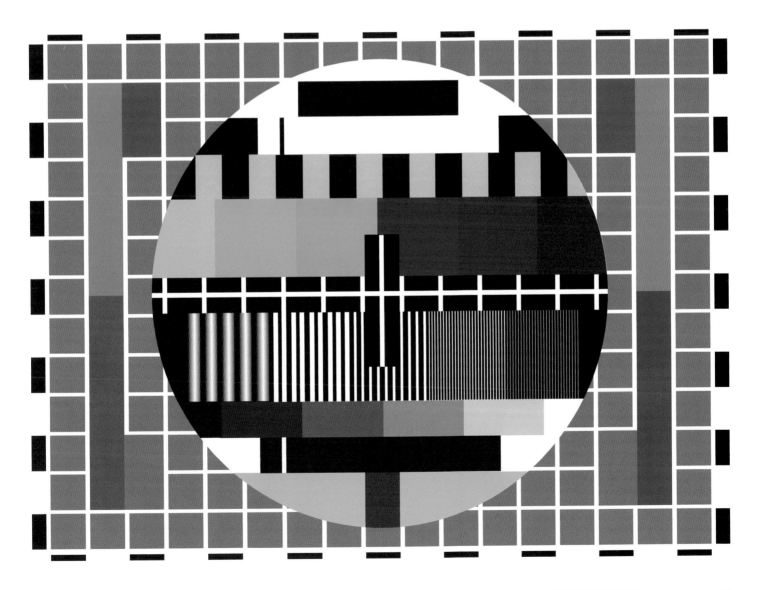

1872
Eadweard Muybridge

CINEMA

The glamour of the modern-day movie industry disguises the simplicity of its roots. A series of photographs capturing the movement of a horse by early photographer Eadweard Muybridge was to pave the way for the moving image, as it exists today. It was Muybridge's approach to photography that allowed the major step from the spinning illustrations of the magic lantern zoetrope to actual photography in motion. The first 'moving picture' commission that Muybridge received was in 1872, which called on him to settle a dispute over whether horses fully left the ground when galloping. A series of cameras sent up alongside the racetrack and triggered in sequence as a horse ran by. Not only did it prove that horses did in fact leave the ground when galloping, but it also gave birth to the moving image.

The next step was the invention of 'kinetograph' in 1891, which was able to take photographs at a rate of 40 images per second. William Dickson, an assistant to Eadweard Muybridge, can be considered responsible for most of its development. In essence it combined the photographic advances made by Muybridge with the technology of the zoetrope, and the subsequent moving series was viewed in a booth called a 'kinetoscope'. Photographic material manufacturer Antoine Lumiere visited one of these booths and immediately instructed his sons, August and Louis, to try and emulate the invention by developing a projection system that could play the film to an audience. With this the 'cinematographe' was born and in March of the same year the first commercial projection of scenes in motion, occurred in the basement of a Parisian café.

By 1914 the focus of the film industry was centred in the USA— more specifically Hollywood, California. Motion pictures remained a silent affair for a long time and were often accompanied by live music, played by an in-house pianist, or occasionally with a specific phonograph record that was made to synchronise with certain scenes in the film.

However, the stars of the silent motion pictures had a rude awakening in 1927 when a film stock was invented that could record both image and sound at the same time. Directors pounced upon the opportunities that this new medium offered, with many musicals following. Sunny Side Up, an early musical that set box office records for the Fox Company in 1929, featured a plethora of accents—Irish, Italian, Swedish and American—as well as background noises such as traffic, dogs and children playing. Traditional visual and slapstick humour was replaced by sophisticated gags using dialogue filled with puns and wisecracks.

The invention of a film that could record both sound and image was initially the bane of the cameraman, who had to adapt to working inside soundproofed, padded boxes so that the microphones did not pick up the whirring of the camera. Furthermore, certain fabrics could not be worn by the actors— in particular tulle, as it rustled so much that it could be heard over the actors' voices. The effects of the introduction of audio stretched so far that cinemas themselves had to be reconstructed to accommodate the use of sound. 'Picture palaces' built with acoustics suitable for live music that previously accompanied

Top An 1882 illustration of a praxinoscope, a device for projecting animations and an early precursor to the zoetrope.

Opposite A selection of photographs by Eadweard Muybridge.

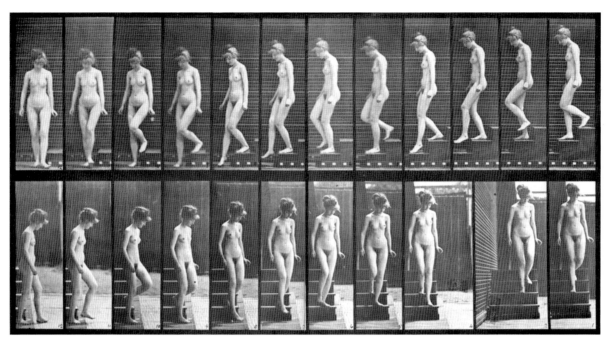

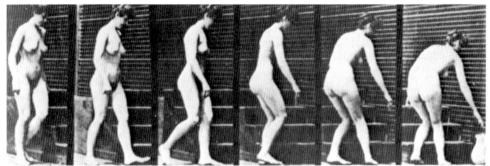

silent movies, had to be modified to minimise reverberation of sound and maximise the clarity of the dialogue.

Colour had been used, albeit frugally, since the first decade of filmmaking, when black-and-white films were sometimes laboriously hand-tinted with colour. The moving picture remained in black-and-white until colour photography began to take a hold after the Second World War. However, a degree of ambivalence towards colour meant that black-and-white dominated on the silver screen until well into the 1960s. British photographer George Albert Smith invented the use of a two-colour process that he called Kinemacolour. More widely renowned, however, is Technicolor, which started life as a dual-colour process until 1930, when the first full Technicolor film was produced. One drawback, however, was that the dyes used in dyeing the film restricted the quality of sound reproduction. As a result full colour was used for less than one film in ten until the late 1940s. Technicolor's technology was unique, recording red, blue and green images on three different strips in a process that required a 'beam splitter' camera that had to be rented by the studios from the Technicolor Company. In 1952, however, Eastman Kodak produced Eastmancolor—a single-film 'monopak' process that did not require a special camera like the 'beam splitter'. TV eventually altogether switched to colour in the 1960s, with Hollywood abandoning the now outdated black and white picture.

The introduction of television saw the number of Americans sitting down in the cinema plummet from 90 million a week in 1948 to 551 million in 1951. This became the catalyst for Hollywood to try and devise something that cinema could provide that could not be replicated at home. One of these systems was 'Cinerama', which used a wide screen to provide a bigger, better and more 'real' viewing experience. The problem though was the cost of the conversion work required to accommodate the three much larger projectors required. The expense left movie houses running the same title for up to two years, trying to recount costs. Three-dimensional systems were much cheaper to install and, although in theory the idea was promising, the special effects were invariably poor and people disliked wearing the uncomfortable glasses. The real breakthrough came in 1952 when Twentieth Century Fox introduced a system called 'cinemascope'. Cinemascope used a lens system, which compressed an image horizontally when photographing it, and then 'stretched it back out when projecting; which gave the effect of objects on screen being part of the audience, its super-wide screen activating the viewers' peripheral vision in a way that resembled normal visual activity.

In recent years, the development of digital technology has had a vast impact on contemporary cinema. Blockbusting cinema has seen the incorporation of special effects created by computer—Computer Generated Images (CGI). Several major Hollywood films have dispensed with film altogether and have been shot entirely in digital format, while other films only use actors' voices for characters, whilst on-screen they are replaced by computer-generated graphics in the emergence of animation cinema. The benefits and disadvantages of this new digital medium continue to be hotly debated, with some major filmmakers moving deliberately in the opposite direction, using older camera techniques for their new productions.

1887
Emile Berliner

GRAMOPHONE

On 8 November 1887, Emile Berliner, a German immigrant working in Washington DC, developed and patented a new system of sound recording. He began working on his machine after seeing the unveiling of Bell and Tainter's Graphophone the previous year. However, Berliner's Gramophone was the very first of its kind, using rotating flat discs with etched grooves rather than using cylinders for the purpose of recording and playing records.

Berliner's discs represented the first recordings with mass-production potential. Berliner had the ingenious idea of using these glass records as a negative matrix in order to produce hundreds of discs from one master copy. The first copies were made out of Zinc; in 1888 it was substituted for celluloid, which tended to wear down too quickly, however, and was replaced in 1889 by vulcanised rubber and in 1896 by shellac (a quite fragile material). Both the discs and Gramophone were produced and distributed by his newly founded 'The Gramophone Company'.

Berliner worked on a clever marketing strategy to promote his Gramophone system, hiring sound engineer Fred Gaisburg to record very talented and famous singers of the time such as Enrico Caruso, Dame Nellie Melba and George J Gaskin. The company averaged out the costs by also recording cheap local talent. It worked and, by late 1894, 1,000 of Berliner's machines and 25,000 records had been sold; the seven-inch, one-sided discs sold at 60 cents each. The second ingenious part of his marketing scheme was the use of the painting *His Master's Voice* by English artist Francis Barraud, which depicted the artist's own dog, Nipper, cocking his ear towards a Gramophone replaying his owner's voice coming from the horn. Berliner gained permission to use the painting as his trademark in 1900 but it was too late to be used for his company and was passed on to Elridge Johnson who he was working with to improve the Gramophone, and to whom he eventually sold the Gramophone system's licensing rights. The trademark was printed on Johnson's Victor Records label, which later conjoined with Columbia records (which in 1948 invented the LP, made from flexible plastic and with the capacity to carry over 20 minutes of sound), monopolising the recording market. Gaining major popularity in the United States, the His Master's Voice trademark was then set to become the world-renowned iconic image that it is today.

STUDIO 1

SOUL JAZZ RECORDS PRESENTS

STUDIO ONE DJ'S

Recorded by Jamaica
Recording Studio,
13, Brentford Road,
Kingston, Jamaica

SJR LP58

Produced by C.S. Dodd

33 RPM

SIDE ONE

1. COUNT MACHUKI
"MORE SCORCHA" (Sound Dimension)

2. PRINCE FRANCIS "ROCK FORT SHOCK" (Scorcher)

3. DENNIS ALCAPONE
"POWER VERSION" (D.Alcapone/C.Dodd)

4. DILLINGER "NATTY KONG FU"
(Dillinger/C.Dodd)

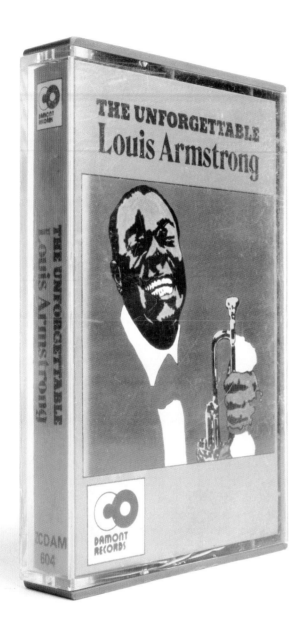

CASSETTE TAPE

The invention of the cassette tape began with developments made in the field of magnetic recording by German inventor Semi Joseph Begun. A sound-recording pioneer, Begun was responsible for inventing the first tape recorder to be used within the media industries. He is also credited with improving the properties of the tape's coating paper and ferromagnetic powders, which when exposed to a magnetic field were permanently magnetised by the field unless erased, allowing the user to record and re-record instantly.

The cassette tape, however, was actually patented and introduced to Europe by Dutch company Philips in 1962. Its invention revolved around the necessity to provide a smaller, compact and portable version than that of the bulky reel-to-reel tape being used previously. The compact audio-cassette used high-quality polyester 1/8-inch tape manufactured by BASF (a chemical and fibre manufacturer), and increased recording and playback times. The next year it was introduced to the United States around the same time that the Norelco Carry-Corder dictation machine was being marketed, which used the blank audio-cassette tape for personal music-recording; the demand for the blank tapes was huge and completely unanticipated by Philips.

1972
Andreas Pavel

PERSONAL STEREO/
WALKMAN

After developing their cassette tape in 1962, Philips made the new technology available free of charge to manufacturers all over the world, based on a patent in 1965. The creation of this device completely changed the way in which people listened to music and would become the analogue predecessor to the iPod.

A German-Brazilian inventor has only recently been credited with the invention of the portable stereo-cassette that was later popularised by Sony as the Walkman. In 1972, Andreas Pavel invented a device that he named the "Stereobelt" and over the following five years he tried to persuade companies such as Philips, Grundig and Yamaha to manufacture it on a mass scale, with no success. However, by 1977, Pavel had patented his invention in Italy, Germany, England, The United States and Japan.

In 1979, Sony created the Sony Walkman TPS-L2 (apparently independently of Pavel's invention but yet very similar in design). It was small, portable, had no record function; it also came with pioneering lightweight headphones for comfort when on the move. Sony popularised the Walkman all around the world, including the countries where Pavel had patented his Soundbelt. In 1980 Sony were forced to begin talks with Pavel, offering him royalty fees but only for sales in Germany. Pavel found the offer unsatisfactory and after many years of legal feud a settlement fee was agreed in 2003. The settlement also gives Pavel recognition from Sony as the original inventor of the Walkman, which was only achieved after the death of the founder of Sony—Akio Morita.

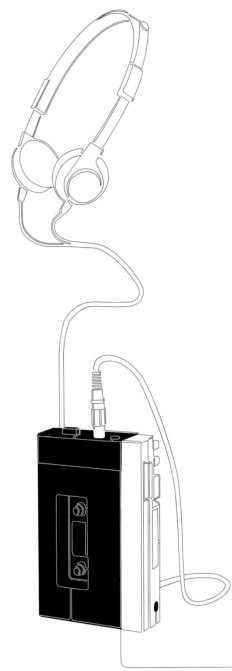

1601447   COMPLETE SPECIFICATION
2 SHEETS   *This drawing is a reproduction of the Original on a reduced scale*
Sheet 2

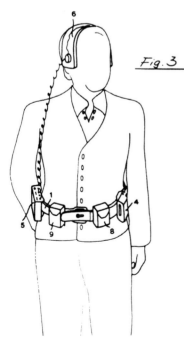

Fig. 3

1965
James T Russell

COMPACT DISC

The compact disc has its origins in the developments made in digital recording by American inventor James T Russell. Russell was dissatisfied with the poor sound quality of vinyl discs caused by the wear and tear of using a stylus, and wanted to find a way to avoid this. In 1965 he was granted a patent for a system of optical recording and playback that involved digitally recording and encoding on to a disc using a very finely-focused laser which was able to read this information optically without damaging the disc. Over the years Russell gained 22 patents for various elements of his system and eventually sold his invention to Sony and Philips.

Although Philips engineers had developed both an optical disc for storing video data in 1969 known as the Video Long Player, and a prototype player made out of glass in 1972, there had been no developments in the advancement of storing audio data on disc. Philips engineer Lou Ottens (who had worked on the cassette tape 15 years earlier), began development on the high-quality audio-disc based on Russell's invention, and in 1979 Sony and Philips formed a joint task force of engineers who developed a whole new system and discs. The discs were 120 mm in diameter and held 74 minutes of audio data, apparently at the insistence of Sony Chief Executive, Akio Morita, who wanted the discs to be able to contain the whole of Beethoven's ninth symphony *Ode to Joy*.

The modern-day compact disc weighs around 16 grams and is 1.2 mm thick. Composed almost entirely of polycarbonate plastic, a thin layer of aluminium is applied on one surface to make it reflective, which is in turn protected by a film of lacquer. The CD data is stored by a series of miniscule indentations known as 'pits' on a tightly-wound spiral track in the layer of reflective aluminium; this is then read by a laser in the CD player which measures the varying degrees of light reflected either as 'pits' or 'lands'.

The first players and discs went on sale in shops in late 1982, the same year as the first album ever to be released on CD: Billy Joel's *52nd Street*. This prompted the digital-audio revolution—the system was enthusiastically received and by 1983 the compact disc had taken over the music market from vinyl discs. In 1985 Dire Straits became the first band to sell a million copies of their album *Brothers in Arms* on CD.

**1977
Fraunhofer-Gesellshaft**

MP3

From as early as 1977 Professor Dieter Seitzer and mathematics and electronics specialist Karlheinz Brandenburg, had been researching methods of compressing audio data. It was not until a decade later, however, that the research centre at German company Fraunhofer-Gesellshaft began to develop technology for audio compression, through a high-quality, low bit-rate audio encoding process. MP3 technology took a few years to reach completion and in reality almost never materialised. Brandenburg explains: "In 1991, the project almost died. During modification tests, the encoding simply did not want to work properly. Two days before submission of the first version of the MP3 codec, we found the compile error." The inventors named on the patent for MP3 technology are Brandenburg, Bernhard Grill, Thomas Sporer, Bernd Kurten and Ernst Eberlein.

The basic premise was to reduce the file in length only up until any deterioration in sound quality occurred: sound can only be heard on a short section of audio range so the software developed would be able to scan and delete any extraneous sound that was inaudible to the human ear. Fraunhofer-Gesellshaft explained: "Without data reduction, digital audio signals typically consist of 16 bit samples recorded at a sampling rate more than twice the actual audio bandwidth (44.1 KHz for CDs). So you end up with more than 1.400 Mbit to represent just one second of stereo music in CD quality. By using MPEG audio coding, you may shrink down the original sound data from a CD by a factor of 12, without losing sound quality."

The first successful MP3 player—the AMP MP3 Playback Engine —was designed by developer Tomislav Uzelac in 1997, Known as Winamp, the free MP3 music player allowed the transfer of high-quality music to people's home PCs via the internet, and began to shape the music industry in its own unique way by allowing small bands to reach a wider audience.

**1931
Alfred Mosher Butts**

SCRABBLE

Alfred Mosher Butts, the inventor of Scrabble, proclaimed: "If there hadn't been any Depression in the 1930s, there wouldn't be any Scrabble." An architect forced out of work during the Great Depression, Butts in 1931 found himself with time to spare and took to devising a board game. He came up with a game of words that included an ingenious mixture of crosswords and anagrams; its success depended in large part on its good balance of skill and luck.

The very first incarnation in the 1930s was called "Lexiko". It had no board, only tiles and was scored on the length of the words created. Its major ingenious trait lay in its letter distribution and the frequency with which individual letters were found within the whole set of tiles. Butts realised that there would be too much luck involved and that games would grind to a halt because of the lack of vowels in relation to consonants, so Lexikon used a collection of letters that were directly reflective of the frequency with which each was used in the actual English language—an idea sourced from Butts' childhood reading of Edgar Allen Poe's *The Gold Bug*. In Poe's short story, a pirate's treasure map is written in code and is deciphered by comparing the frequency by which each mysterious symbol is used with the frequency each letter is used in the English language. Butts quantified the frequency with which each letter was used by tallying up their occurrence from the pages of newspapers. Vowels, especially 'e', were extremely common; 'q' and 'z' were least; thus, infrequent letters became worth more points and vice-versa.

The game changed its name to "It" and "Criss-Cross Words" after being rejected by numerous game and toy companies, even with the addition of a board. It was not until 1948, when entrepreneur James Brunot took interest, that its commercial potential was seriously considered. Brunot proceeded to make improvements to its design and name, deciding on the now world-famous moniker "Scrabble". Even with these improvements, sales were slow and remained that way until 1952 when the owner of Macy's department store played the game whilst on holiday and on his return instructed the toy department to stock the board game. Many other toy and game retailers followed suit soon after.

Eventually, New York firm Selchow & Righter, who were making the boards for Brunot, bought the rights for the whole game. Readers of *Games and Puzzles* magazine voted it as Game of the Year in 1975. Today Mattel owns the rights to Scrabble in most countries. Over 100 million Scrabble sets have been sold to date, in 29 different languages, each taking into account the particulars of each language with respect to letter distribution.

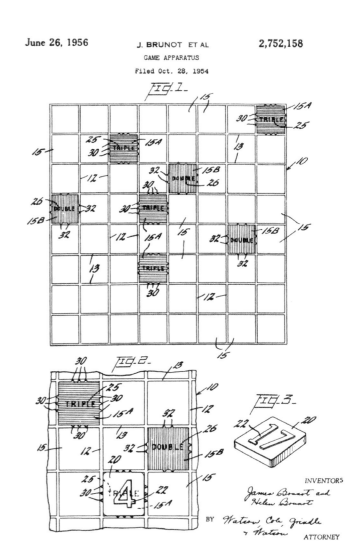

MONOPOLY

During the American Depression in the 1930s, Charles Darrow devised a board game—today's Monopoly—with Atlantic City's street names painted on a tablecloth and houses and hotels made out of scrap paper. Darrow invited his friends round to play his game, after which they all asked to buy sets, which he sold for $4 a piece. Darrow wrote to games company Parker Brothers about his board game after he was unable to keep up with the demand for the game—which he was now selling at a more expensive $10—but the game was rejected out of hand for having 52 fundamental errors, one of them being that it was too long and complicated. However, the wife of Robert Barton, the president of Parker Brothers, convinced him to buy a set, a move that saw him have a change of heart and buy the rights to the game, making Monopoly a registered trademark in America in 1935 and in Britain in 1952. Monopoly became internationally popular and the streets of Atlantic City were swapped for the streets of London, Paris and Rome, amongst others. Darrow became the first games inventor to become a millionaire and a plaque was erected in his honour in Atlantic City. It is estimated that over 500 million people have played the game around the world.

There is, however, another story of how the game of Monopoly came to be. Games that involved the buying or selling of shares or property gained popularity during the beginning of the twentieth century, many of which shared a close resemblance to Monopoly. One such game was "The Landlord's Game", invented by Lizzie Magie from Maryland, USA. The board game had no street names but did have the price of rent and sale on each of its squares. Whilst Darrow's game allowed players to revel in a fantasy world of capitalism in the midst of a depression, Magie—a staunch supporter of the radical political economist Henry George—meant for the game to be an attack on capitalism. It is then believed that a man called Dan Layman from Indiana, USA, got to hear of the game and reworked its rules and game-play into a version called "Finance". This game had additional features attached to it by a chain of people, including the addition of the street names of Atlantic City by a local resident. It is apparently at this point that the game was introduced to Darrow.

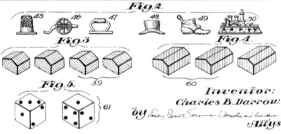

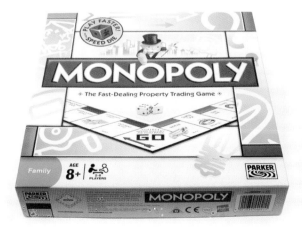

CHESS

Problems occur when trying to pinpoint the exact origins of the game of chess. The modern game as we know it is a mixture of ideas and practices from different cultures acquired along early nomadic trade routes, such as the Silk Road. It is, however, highly probable that one of the very first incarnations of the game was a sixth-century Indian war game known as *Chaturanga* ("eight legs" in Sanskrit). The game spread throughout the world, first to Asia, Persia and then to Europe.

Another version of the game—*Shatranj*—hit Western European shores via the Moorish invasion of Spain in the eighth century, and on its journey over the next 900 years evolved into the basic modern game that is played today. The most noticeable developments in the rules occurred during the fifteenth century. The three main modifications were that each pawn was permitted an extended move on its initial turn; the range of the bishop was extended; and the role of the queen was drastically changed by promotion of power from the relatively weak queen of the medieval age to the most effective piece on the board, able to move both as bishop and rook. This changed the strategies needed to play and win chess.

By the end of the Renaissance period the Italian Gioachino Greco dominated the chess world, and travelled around Europe relating his ideas and theories about the game. The French middle-class became enthralled and Paris became the centre of chess; clubs were set up and top players met up at the Café de la Régence to play each other. Then in 1843 Englishman Howard Staunton beat French champion Pierre Charles de Saint-Amant to become chess master, and subsequently London became the new capital of chess. Staunton also went on to devise a system of notification to both record games and teach, as well as creating a standard design for chess pieces.

Modern-day professional chess players not only have to face competition from human competitors but also from the silicon world of computers; the precursor to these computer chess players being "The Turk" constructed by Wolfgang von Kempelen in 1770. The Turk was an automaton that was paraded around Europe, bamboozling and beating many great chess players such as Napoleon and Benjamin Franklin. For 80 years the machine bemused and entertained spectators moving pieces of its own accord, and apparently by conscious effort. Finally it was revealed that The Turk was an elaborate hoax and actually operated by a chess player concealed in the cabinet of The Turk with the use of a secret door and mirrors. In effect it was the first great cabinet illusion in the history of magic. The first actual chess computer was built in 1958 and by 1974 the first computer chess world championship was held. As the amount of information computers could hold increased so did their playing ability, changing the nature of chess altogether. The first computer to challenge a chess master was Deep Blue; Gary Kasparov, the reigning world champion at the time, beat Deep Blue in 1996 but then lost a rematch the following year. Since then a team of Grandmasters have been programming the hardware with every possible endgame position with five or fewer pieces.

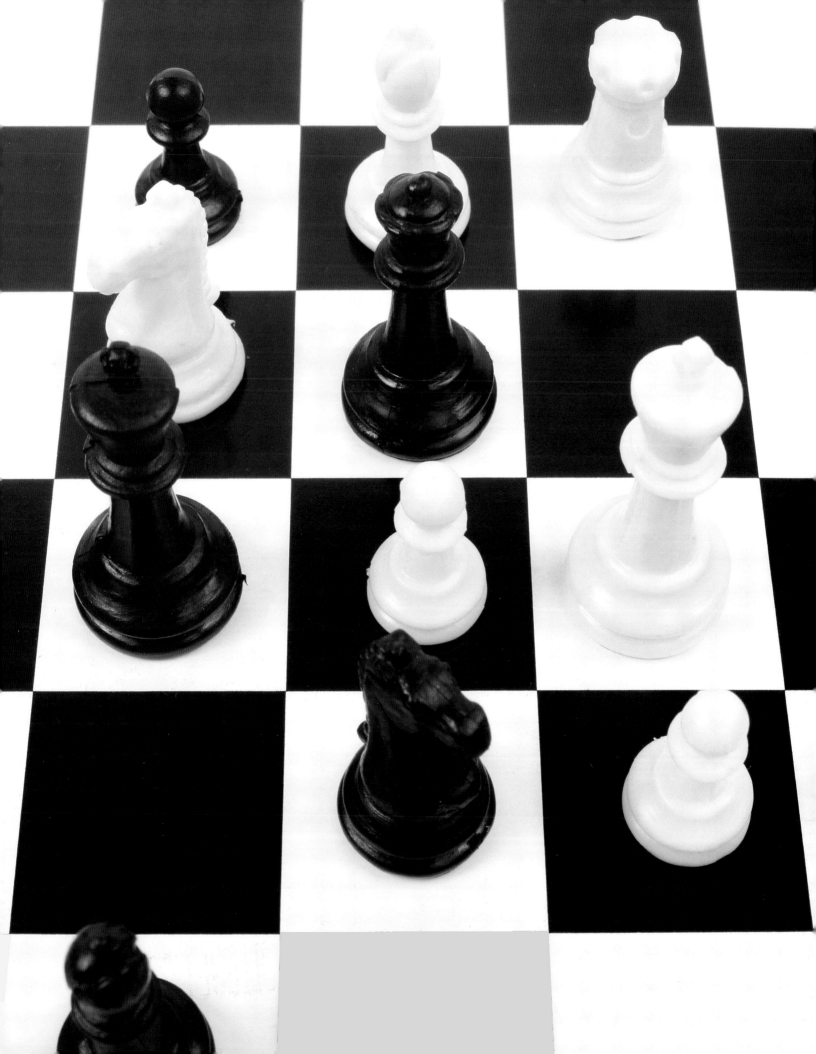

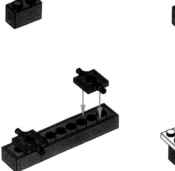

1932
Ole Kirk Christiansen

LEGO

The company LEGO was founded by a Danish joiner in 1932, Ole Kirk Christiansen, who went door-to-door selling stepladders, ironing boards and wooden toys that the company made. The name "LEGO" was the amalgamation of two Danish words "leg" and "godt", which translates as "play well".

With the development of plastics after the Second World War, LEGO began to make "automatic binding bricks" in 1949 after becoming the first company in Denmark to buy an injection-moulding machine. These bricks had the now distinctive studs at the top of the brick but not the hollow cylinders for them to fit into. This was soon to change, however, after Ole's son attended a local toy fair in 1954 where buyers complained that all the toys on show seemed to be the same. Ole then designed the hollow cylinders that would connect with the studs, which would afford increased stability and provide a design that allowed compatibility between older and newer lines of the same product. The brick, along with the interlocking principle, was launched in 1958.

After dropping wooden toys altogether in 1960, LEGO began churning out their popular plastic building bricks and the variety of accessories that came with them. First roofs, floors and wheels, and then trains in 1966. The following year saw the release of DUPLO—a game featuring bigger blocks that could be used by infants. In 1977, the TECHNIC line of sophisticated projects for older kids and teens was marketed and, by 1997, blocks that included software that could be programmed to carry out certain functions called 'Mindstorms' were introduced at the Museum of Science and Industry in Chicago, USA. By 1998 there were 600 different sets based around eight main lines including pirate, space, and castle sets. LEGO has now become part of movie franchising with all major film series, such as Star Wars and Indiana Jones, having their own LEGO line.

LEGO has become one of the major Danish exporters and it is estimated that 70 per cent of American and 80 per cent of European households own LEGO. Its popularity eventually led to a Legoland—a theme park that is almost entirely made out of LEGO—being built in Billund, Germany, with others shortly following in Windsor, England and California, USA.

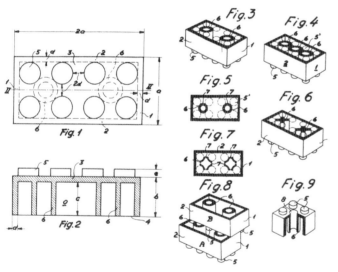

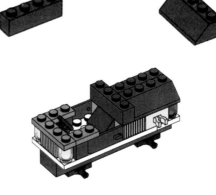

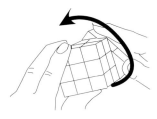

**1974
Ernö Rubik**

RUBIK'S CUBE

Invented by Hungarian sculptor and inventor Ernö Rubik in 1974, the Rubik's Cube is considered to be the world's best-selling toy, with more than 300,000,000 having flown off the shelves since it first went into production under that name in 1980. The classic model measures just under six centimetres on each side and consists of 26 miniature cubes, their outward-facing sides featuring one of six colours: traditionally white, yellow, orange, red, blue and green. A pivot mechanism in the centre allows the faces of the cube to turn independently, thus creating a multitude of potential colour combinations: 43,252,003,274,489,856,000 to be exact. The puzzle is solved once each face of the cube is one solid colour. Solutions for the Rubik's Cube vary and include different methods such as solving faces or corners independently.

Since it first appeared on the market, people have entered speed-cubing competitions, testing their dexterity in an attempt to solve the cube in the shortest time. Such competitions are held on an international scale and the competitor has five attempts—from which an average time is recorded, as well as the time of their single best attempt. Erik Akkersdijk, who managed to solve the Cube in 7.08 seconds at the Czech Open, holds the current 2008 world record. A number of alternative Rubik's Cube competitions have also emerged throughout the toy's history, the most interesting of which include solving the puzzle blindfolded, using a single hand or one's feet, or even solving it underwater in a single breath; the latter, unsurprisingly, have been refused an endorsement by the World Cube Association.

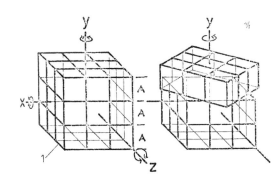

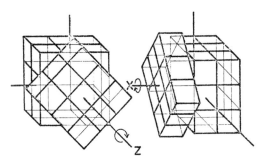

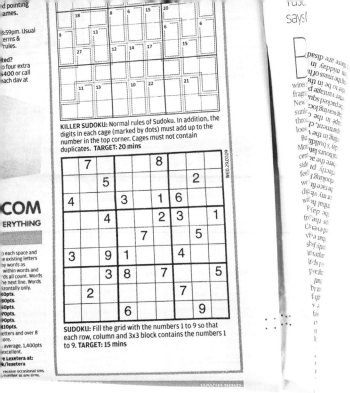

KILLER SUDOKU: Normal rules of Sudoku. In addition, the digits in each cage (marked by dots) must add up to the number in the top corner. Cages must not contain duplicates. **TARGET: 20 mins**

SUDOKU: Fill the grid with the numbers 1 to 9 so that each row, column and 3x3 block contains the numbers 1 to 9. **TARGET: 15 mins**

1979
Howard Garns

SUDOKU

Leonhard Euler, a Swiss, eighteenth century mathematical genius invented the numerical brainteaser that he termed as "Latin Squares" in 1793, the forerunner for the modern-day game of Sudoku. The life of Sudoko as it exists today is, however, much shorter, and its invention is accredited to an American freelance puzzle creator called Howard Garns, who saw his 'Number Place' puzzle published in 1979 in a magazine called *Dell Pencil Puzzles*.

It was not until 1984 that a Japanese publisher made a few adaptations and logistical improvements to the number puzzle and renamed it Sudoku; Su meaning number, and Doku meaning singular or unique: Singular Number. It became hugely popular in Japan and by 2004 it was picked up by British newspaper, *The Times*.

The modern-day Sudoku puzzle comprises a nine by nine grid, which is divided into nine smaller boxes of three by three. The aim is to fill all cells in the grid without having the same number in the same row or column, while also making sure that each number appears only once in each of the three by three squares. Sudoku demands a systematic and logical approach. Aids and clues are provided by the allowance of a few numbers to start you off and to make sure that there can only be one possible solution. The fewer numbers there are, the harder the puzzle is to solve, and for this reason the puzzle becomes easier as you move closer to completion.

1913
Arthur Wynne

CROSSWORD PUZZLE

An English expatriate living in America, working for the *New York World* newspaper, devised the crossword puzzle—originally called "the word-cross puzzle". It was 1913 when Arthur Wynne was asked by his editor to invent a new game for the newspaper's 'Fun' section. Wynne apparently recalled a game from his early childhood entitled "Magic Squares", which involved arranging a certain group of words in a particular order so that they would read the same horizontally as vertically. This basic design was increased in size and complexity and included clues to the particular words rather than just the words themselves. This first diamond-shaped crossword appeared in the December 21 issue under the subtitle of "mental exercises". Although there were no black squares, the clues were easy and the answers short in comparison to modern day crossword puzzles.

The puzzle was a huge success and inspired many newspapers, both American and British, to follow suit. By 1924 the first book of crossword puzzles was published by New York's Richard Simon and Max Schuster, selling all 36,000 copies of its first printing. This led to two sequels as well as the move by numerous other publishing houses to release their own versions in hope of replicating the same success.

As time went on the puzzles evolved to become increasingly complex and varied, with multiple-word answers, pun-based and cryptic clues and different grids featuring fewer checked letters. It could be argued that the crossword is the world's most popular word game and that its invention has provided us with an intellectual way of passing time, enriching our vocabulary and capacity for critical thought.

1000 BC
China

YO-YO

The yo-yo has a long history. Having been in existence for over 3,000 years, it is thought to be the second oldest toy still available—with the doll taking the top spot. However, its origins are somewhat vague. It is thought to have first originated in China around 1000 BC, but there are also arguments that it originated in Greece where there is evidence of yo-yos being made from terracotta, decorated with images of Gods and used as toys for young children who then gave them up once reaching adulthood as a rite of passage. Similarly, in the Philippines the yo-yo had a long history as a popular toy for young children, but there is also a mythic story that modified versions were used as instruments of war for over 400 years and were enlarged with razor-sharp edges and studs and attached to ropes up to 20 feet in length for throwing at enemy forces. The now commonplace term "yo-yo" is taken from the Philippine native language Tagalog, and translates as 'come back' or "to return".

The toy became popular during the late eighteenth century in France and was commonly known as a "bandalore", alongside other names that were directly linked to the French Revolution such as *l'emigrette*—meaning "to leave the country" (rather than face the guillotine). The toys picked up the name "quizzes" in England as well as the "Prince of Wales" toy.

The bandalore was used in America around the 1860s but it was not until the 1920s that the toy became commonly known as a yo-yo. Pedro Flores, a Filipino immigrant living in America, began mass-manufacture of the toy in California in 1928. It was then popularised on an immense scale by Donald F Duncan who bought the rights from Flores and then trademarked the name "Yo-Yo". Duncan was a marketing genius and blitzed the media with promotion; at the pinnacle of the fad his factory was producing somewhere in the range of 3,600 yo-yos an hour, while a single promotion in Philadelphia sold 3 million units of the toy in one month in 1931. However, yo-yo sales were as up and down as the actual toy and just three years later Duncan lost the trademark Yo-Yo and went bankrupt.

From a design point of view, Duncan's Yo-Yo model had some novel adaptations that improved its performance and popularity. The standard or imperial shape was reversed to that of a butterfly shape, and the design allowed the player to catch the yo-yo on the string with increased ease. It was also the first yo-yo design whereby the string was not tied to the axle but looped around it, allowing the yo-yo to stop or 'sleep' at the end of the string for the first time and provide it with the capability of performing tricks. During the 1970s, yo-yos were modified in various ways including rim-weighting, which led to a longer spin, and a patent by Tom Kuhn for a yo-yo that could be taken apart to allow for replacement of the axle, called the "No Jive 3-in-1", in 1978. By 1980 Michael Caffrey had patented 'The Brain', which was a yo-yo with a centrifugal spring-loaded clutch mechanism that caused the yo-yo to automatically return to the users' hand once its rotational spin had decreased to a certain speed; and during the 1990s yo-yos were fitted with ball-bearing axles, again increasing spin-time.

FRISBEE

Prevailing stories suggest that the origins of the modern-day Frisbee can be traced to a late-nineteenth century baking company in Bridgeport, Connecticut. The Frisbie Baking Company had an ingenious advertising tool whereby the words "Frisbie's Pies" were embossed on to the bottom of the tin pie plates. These empty pie plates were adopted for recreational purposes in the surrounding community with many New England Colleges incorporating the empty tins into a new student sport—the participants yelling "Frisbie" as they did so.

The sport grew in popularity during the Depression and was spread across the United States by soldiers during the war. The game, however, did have some drawbacks: in the centre of the tins were six perforated holes in the shape of a diamond, which made a displeasing shrill noise when thrown; on top of this, after perpetual crashing they would either crack or develop sharp edges that were capable of cutting fingers. So, in Los Angeles, in 1948, two Second World War veterans, Walter Fredrick Morrison (the father of whom invented the automotive sealed-beam headlight) and Warren Franscioni had the idea of casting a Frisbee design in plastic, and changing the shape so that it had a sloping edge. Morrison and Franscioni's partnership dissolved but Morrison persisted with their idea and gained the patent for the 'Morrison Slope'. The slope changed the aerodynamics of the design, making its flight smoother and lengthening the time it stayed afloat.

Morrison named his new version the "Pluto Platter" because of its flying saucer-like appearance and to cash in on the UFO craze in America that had around the same time been fuelled by incidents such as Roswell in 1947. Morrison frequently gave demonstrations of the Pluto Platter, showing people how to throw the object correctly. Many observers thought that the disc followed an invisible wire having not seen anything travel like it before. In 1955, two of these observers were Rich Knerr and Spud Melin, who owned a toy company called Wham-O. Morrison signed a contract with them, and Knerr and Melin marketed the flying object with expertise, renaming it the "Frisbee", starting production in 1957. Sales rocketed due to the Frisbee being publicised as a sport; which would eventually develop into Ultimate Frisbee in the late 1960s, using professional Frisbees first designed by Ed Headrick for Wham-O in 1964. This Frisbee incorporated concentric grooved rings, called the "Rings of Headrick", which stabilised its flight and allowed the Frisbee to travel for longer and with greater accuracy. It was not until the 1990s that the concept of the Frisbee would undergo another major development, this time in the form of the Aerobee, a rubber ring that was so light it could cover previously unimaginable distances.

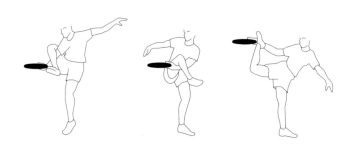

Fourteenth Century

HULA HOOP

More than 3,000 years ago, children in Egypt played with large hoops of dried grapevines that they propelled along the ground with a stick or swung around at the waist. During the fourteenth century, a 'hooping' craze swept England, popular with both adults and kids; the records of doctors at the time attribute numerous dislocated backs and heart attacks to hooping. The word "hula" became associated with the toy in the early 1800s when British sailors visited the Hawaiian Islands and noted the similarity between hooping and hula dancing. In 1957, an Australian company began making wood rings for sale in retail stores.

The item eventually attracted the attention of Wham-O, a fledgling California toy manufacturer. Richard Knerr and Arthur Melin, of Wham-O, manufactured a plastic hoop in a variety of bright colours. Knerr and Melin promoted it for months in 1958 on Southern California playgrounds, where they would do demonstrations and give away hoops to get the children to learn and play. Their perseverance turned HULA HOOP into the greatest fad the country had ever seen, with a staggering 25 million being sold in just two months! The fad eventually died out in the 60s, but Hula Hoops are now very much with us again thanks to the current fitness craze.

ICE SKATES

During an archaeological dig in the Bjoko region of Sweden, a pair of ice skates were unearthed that appeared to date from as early as 1000 BC. Made from animal bone—usually the tibias (leg bones) or jawbones of reindeer, elk and oxen, the bones had a hole bored into each end through which a strip of animal skin passed to be tied around the foot. Interestingly, an old Dutch word for skate is "schenkel" which means 'leg bone'. These skates enabled easier movement over hostile terrain and were used in hunting; given the material from which these skates were made, they were frequently brittle and invariably hard to manoeuvre as they glided on top of the ice rather than skating on it. As a consequence, poles were often also used in order to steady the wearer.

It was not until the fourteenth century that advancements in skate technology in Holland saw the addition of steel bottom-runners attached to wooden platforms that tied to the shoe via leather straps. The steel runner blade was double-edged and sharpened, which meant that the skate actually cut into the ice, aiding movement by allowing the skater to push and manoeuvre with their feet. This was called the "Dutch Roll" and made the use of a pole unnecessary and redundant. It was around this time that the use of ice skates in leisure and sport radically overtook their use practically.

The use of skates for recreation exploded with the refinement of ice skate design. In 1848, EV Bushnell of Philadelphia developed the first 100 per cent steel clamp and, in 1865, a famous American skater, Jackson Haines, invented an all-metal blade that attached directly to a boot without straps; he also added the first toe-pick to a skate blade in the 1870s. The basic design of the ice skate has not changed in any major way since then; however, three different types of ice skate have evolved with their own specific traits to allow the user to perform to the best of his or her ability in their particular sphere—the hockey skate, figure skate, and speed skate.

1760
Joseph Merlin

ROLLER SKATES

Joseph Merlin, a Belgian-born, London-based inventor, first demonstrated a pair of rudimentary roller or 'dry-land' skates in 1760; they consisted of a small row of metal wheels in a line, attached to a boot. After testing his prototype out on his 13 year old grandson, Merlin thought he would demonstrate the invention himself by famously wearing them to a masquerade party. Hoping to create a grand entrance, Merlin rolled into the ballroom playing a violin and promptly sped into a very expensive, wall-length mirror hanging on the opposite side of the room, smashing it—and his violin—to pieces. Stopping and manoeuvrability were two obstacles of design that Merlin failed to overcome in his roller-skating prototype.

The gradual development of roller skates meant that in 1818 they made a more elegant entrance onto the public stage, quite literally this time, in the German ballet *Winter Pleasures*. The ballet was intended to be performed on ice but, as there was no way of reproducing ice on a stage at the time, roller skates were used as a substitute. It was also in this year that the first patent for a roller skate was granted to French inventor Monsieur Petibledin, with inline wheels made of copper, ivory or wood.

The birth of the modern four-wheel skate came via James Plimpton's invention in 1863. The American's roller skates consisted of two parallel sets of wheels located at the heel and ball of the foot and were mounted on wooden springs, which enabled greater manoeuvrability and allowed travel in curved lines. Following the invention of the ball bearing and the advent of plastics, the 1960s, 70s and early 80s saw roller-skating's popularity grow, with the sport eventually teaming up with the world of disco music—leading to the opening of some 4,000 roller-discos.

The Rollerblade skate arrived on the scene in 1979 via the aid of two hockey-playing brothers, Scott and Brennan Olson. The Rollerblade was based on the first inline roller skate designs but were improved exponentially by using modern materials such as polyurethane wheels attached to ice-hockey boots, and the addition of a toe break. After ironing out some major flaws and continuing to make improvements in design as technology progressed. Rollerblades became a massive success and remain so today.

SKATEBOARD

The skateboard was originally conceived as a surfing replacement in the absence of a nearby coastline or appropriate weather conditions. During the late 1950s, frustrated surfers in California attached the wheels from roller skates to wooden planks to allow 'sidewalk surfing' when coastal conditions were unsatisfactory. What began as a local trend quickly caught on throughout the major cities of the US.

By the 1970s, manufacturers responded to demand and began producing boards with increased strength, flexibility and stability, opening opportunities up to the users, who could perform tricks with increasing dexterity. Moves such as the 180, the 360 and the famous 'Ollie' soon became commonplace manoeuvres. The most well known trick—the 'Ollie'—was invented by Alan Ollie Gelfand, and involved flicking the board off the ground while keeping contact with the feet.

Large industrial spaces and construction sites naturally attracted skateboarders with their freely available 'ramps' in the form of giant half pipes or empty swimming pools with gradually sloping sides, helping to add to the identity of the emergent culture. Over time, these sites were emulated in the form of official skate parks featuring ubiquitous obstacles.

ABSOLUTELY
GO
SKATEBOARDING
ON THESE
PREMISES

ROLLERCOASTER

The first fully operational roller coasters were used in France in the early nineteenth century. The two roller coasters—one in Paris and one in Belleville—were the first to have wooden cars connected and locked to a metal track by a wheel axis and they would fly down a single slope (after being pushed up the hill by the attendants) at speeds of 30 mph. Although the roller coasters initially provided a new form of entertainment for the French population, the novelty soon wore off and it was at this point that a 'Gravity Rail Road' was unveiled in Pennsylvania USA. The Mauch Chunk Railroad was a disused coal train that was pulled up to the top of a slope by a coal-powered engine; at the top, the train was sent speeding back down to the bottom of the hill. It was in this spectacle that millionaire La Marcus Thompson could see potential.

Thompson built the first working wooden coaster in Coney Island in 1884. This 'Big Dipper' would travel up and down various slopes in a straight line; however tame this may seem to a present-day audience, this roller coaster was considered incredibly exciting. It was, however, a young 19 year old working at Thompson's company who came up with the innovative design that became the blueprint for all modern-day roller coasters. John Miller devised a system to improve the ride safety of the Big Dipper, ensuring that the wheels of the car never left the track. It involved three sets of wheels, one set above the track to carry the weight of the car, one residing within the track to act as a guide around the course, and one underneath, locking the car to the track. Miller soon realised that by devising a system that made the roller coaster safer to ride he had also realised the potential to make the ride faster and more thrilling by adding twists and turns to the track, and overall creating the illusion of real danger within a system of increased safety. The manufacture of such roller coasters began during the 1920s, giving thrill-seekers an incredibly exciting ride.

United States Patent [19]
Wilson et al.

[11] **4,373,417**
[45] **Feb. 15, 1983**

[54] **ELECTRIC GUITAR**

[75] Inventors: **Gregg Wilson**, Aurora, Colo.; **John F. Page**, La Mirada, Calif.

[73] Assignee: **CBS Inc.**, New York, N.Y.

[21] Appl. No.: **272,198**

[22] Filed: **Jun. 10, 1981**

[51] Int. Cl.³ **G10H 3/00;** G10D 1/08; G10D 3/04; G10D 3/18
[52] U.S. Cl. **84/1.16;** 84/267; 84/298; 84/307; 84/328
[58] Field of Search 84/267, 291, 292, 298, 84/299, 307, 308, 328, 177, 116

[56] **References Cited**
U.S. PATENT DOCUMENTS

2,972,923	2/1961	Fender	84/307
2,976,755	3/1961	Fender	84/1.16
3,290,980	12/1966	Fender	84/307
3,427,916	2/1969	Fender	84/267

OTHER PUBLICATIONS
Fender Brochure, Jan., 1982.

Primary Examiner—L. T. Hix
Assistant Examiner—Thomas H. Tarcza
Attorney, Agent, or Firm—Gausewitz, Carr, Rothenberg & Edwards

[57] **ABSTRACT**

The economy of manufacture of electric guitars and electric bass guitars is improved, with no loss of quality, by providing an anchor flange in integral relationship with a metal pickguard of the guitar or bass guitar. Extended through the anchor flange are adjustment screws which connect adjustably to bridge barrels over which the strings extend. The adjustment screws and bridge barrels are preassembled to the anchor flange, and all electric components are preassembled to the pickguard, prior to mounting of the pickguard on the body of the guitar or bass. Thus, the ultimate in economy is achieved, yet the anchor flange has very strong support from the pickguard and is located accurately thereby.

10 Claims, 4 Drawing Figures

1930s

ELECTRIC GUITAR

Although the electric guitar did not appear until the 1930s, the wish to increase the volume of the classical guitar existed long before the advent of amplifiers and speakers. The nineteenth century experienced musical performances that were characterised by an ever-increasing need for bigger concert venues and ensembles. Musicians needed louder and more powerful instruments, that would perform more efficiently in such surroundings.

In the nineteenth century steel strings were adopted; this facilitated greater volume and also greater tension on instruments. The conventional flattop guitar began to alter in size and shape as a steel-string instrument. Furthermore, an entirely different design appeared, the stronger—and louder—'archtop'.

It was in the late 1930s that electronic amplification paved the way for building a louder guitar, despite the resistance of some traditionalists. Country and jazz musicians were among the first to champion the electric sound. Then throughout the following two decades, players and instrument makers began making Spanish-style electric guitars with solid wooden bodies, leading to new designs and new sounds.

During the 1930s and 40s many important artists pioneered new ways of playing the guitar such as Eddie Durham and Oscar Moore, country players Noel Boggs and Merle Travis, and blues masters T-Bone Walker and Muddy Waters. Their music was marked by an experimentation with the guitar's tonal and harmonic possibilities. Gradually, an ever-increasing array of other artists and audiences started to pay attention to the new electric sound.

Charlie Christian was the first artist to develop a style unique to the electric guitar in 1939. At the same time, a few other artists began experimenting with a new kind of electric guitar; like earlier designs they used the same pickup but mounted the pickup on a solid block of wood. Les Paul, who was already a famous acoustic guitarist, built such a guitar on a four-by-four piece of pine and called it "The Log". Leo Fender, a former radio repairman, introduced a mass-produced solid-body electric guitar in 1950, and Gibson introduced a model endorsed by Les Paul himself in 1952. The solid-body guitars didn't have the feedback problems that characterised hollow-body electric guitars, and they had greater sustain.

Since then, every generation has found a surprising new way of playing the instrument. By all accounts, the electric guitar's impact in contemporary music has been immense.

MOOG SYNTHESISER

Robert Arthur Moog's interest in electronic music systems began when he first manufactured and sold vacuum tube Theremins kits as a student in the early 1950s. However, it was in the 1960s that he began to market more complex systems, and the huge amount of interest that they generated allowed him to set himself up as a company.

At the time, studios producing electronic music were fitted with oscillators, filters and various other sound modulating devices and relied on recording each section on separate pieces of magnetic tape to bring the music to life. It was Moog's invention of the transistor that introduced the ability to build cheaper systems with mass-production potential. A performance using a prototype Moog at the Audio Engineering Society convention in October 1964 saw the synthesizer receive much praise, with Bob receiving orders on the spot.

Moog's system worked by analysing and systematising the production of electronically-generated sounds, breaking down the process into a number of functional blocks and using a standardised scale of voltages for the electrical signals that controlled the functions of the modules. A voltage-controlled oscillator generated the primary sound signal, which could be emitted in various waveforms such as square and sine waves. The signal could be fed back into other voltage-controlled modulators and filters, to create a wide spectrum of sound.

The synthesizer's output was designed so that a keyboard could control it; however, the Moog modular systems were initially intended for high-end studio work and not for live performance. Moog's first customised modular systems were built during 1965 and demonstrated at a summer workshop at Moog's Trumansburg, New York, factory in August 1965; culminating with an afternoon concert of electronic music on 28 August.

In 1967 Moog introduced its first production model, the 900 series, which was promoted with a free demonstration record composed by Wendy Carlos. Carlos began to translate pieces by Bach into the new Moog system, releasing an album *Switched on Bach* on Columbia Records in 1968. Moog played one of these pieces at the AES convention in 1968 and received a standing ovation.

When demonstrated at the Monterey International Pop Festival in June 1967, the Moog began its ascent into the mainstream music world. Amongst others, The Beatles, The Doors, The Byrds and Simon and Garfunkel were fans of the Moog Synthesizer. It is now a legendary piece of kit with great retro value. A number of software versions of the Moog Synthesizer have been made, none, however, quite matching the unique sound of the original machine.

1943
Richard James

SLINKY

Consisting of a simple coil design, originally made of metal but now also produced in plastic, the Slinky was invented in 1943 by Richard James. Given the title of Official State Toy of Pennsylvania, the home state of its inventor, the Slinky has remained a hugely popular children's toy—from the day it was first unveiled in a department store in Philadelphia in 1945, when it sold out in 90 minutes, to today.

   Over a quarter of a billion Slinkys have been sold worldwide since it first went into production. The simple concept was born when James observed a tension spring fall off a table while he was working on a ship and noticed the marketable entertainment value. James' wife Betty decided on the name Slinky after she discovered that it also meant 'graceful' in Swedish.

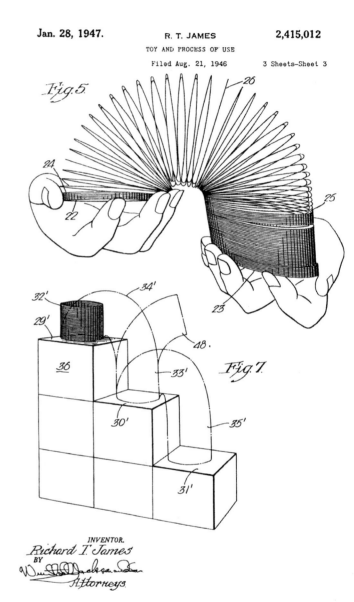

VIEW MASTER

In 1938 a German organ maker and keen photographer called William Gruber envisioned a device that would allow the viewing of three-dimensional images by looking at a slide reel through two eyepieces that consisted of two slightly different images of the same object overlapped. This device, which later became known as the View Master, was based upon the ideas of Sir Charles Wheatstone—published in a paper regarding 'binocular vision' in 1838—and his invention of a device he coined the Stereoscope. Wheatstone succeeded in explaining why we see life as three-dimensional rather than flat, as in a photograph: the depth and perspective of vision is produced by the combination of two slightly different images of each individual eye by the brain, forming a composite image.

Gruber devised a way of using cinema film to create stereo pictures, as well as a way of viewing these images inexpensively. A simple reel holding seven pairs of images taken on 16 mm film, each mounted two-and-a-half inches apart (around the average equivalent spacing of a pair of human eyes), were encased in 'boxed-binoculars'. When the View Master was pointed towards the light a 3-D image became visible, and with a push of a lever the reel would spin, allowing a new pair of pictures, and therefore another full 3-D image, to be viewed.

Lack of funds meant that Gruber could not market his invention, and it was only by a stroke of luck that in the summer of 1938 Gruber met fellow photography enthusiast Harold Graves while visiting the Oregon Caves National Monument, where Gruber was using his 3-D camera. Graves was the president of postcard company Sawyer's Photographic Services and he and Gruber shared an interest in stereography. He was trusted with Gruber's unpatented idea and turned the View Master into a reality. It was marketed as a souvenir for adults in order to view photographs of landscapes and monuments of national parks. It was hugely successful and became Sawyer's main product line. In 1951 Sawyer's bought out the View Master's main competitor—Tru-Vue—therefore owning the rights to produce slides of the cartoon characters of Walt Disney that True-Vue previously licensed. This evolution into the sphere of the children's toy industry saw business boom, so much so that the View Master is now owned by Fisher-Price.

Although becoming a cult toy, the View Master has its place in pioneering science. Gruber and David L Bassett— an expert in anatomy and dissection—began a 17 year project to create an incredibly detailed set of images of the human body, both internally and externally, that could be viewed in 3-D. The *Stereoscopic Atlas of Human Anatomy* was published in 1962. It consisted of 25 volumes, including 1,500 pairs of slides, that produced 3-D images of formaldehyde-injected cadavers when looked at through the View Master.

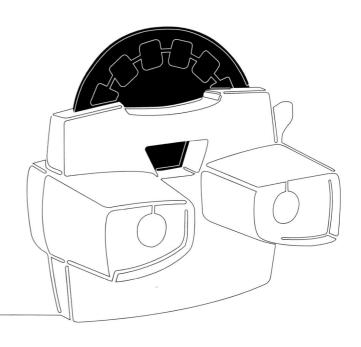

Late 1950s
André Cassagnes

ETCH A SKETCH

During the late 1950s, a Frenchman by the name of André Cassagnes, developed a drawing toy—the "Telecran"—using aluminium powder, a glass screen and a joystick. It was spotted by the Ohio Art Company and with the aid of one year's research and development it gained its classic design, shape, and colour as well as its new name—the Etch A Sketch. It became an overnight sensation during the Christmas season of 1960; the factory was so overwhelmed with demand that they continued to manufacture the product right up to Christmas Eve in an attempt to meet with customer orders.

The design of the Etch A Sketch has hardly changed since its first conception, with only minor alterations made to the two knobs in order to allow for easier grip. New pink and blue variants were released during the 1970s but have not attained the popularity of the classic bright red model. The one thing that has remained exactly the same is the inner working of the drawing tool. The red frame encases a glass screen, on the reverse side of which is a coating of extremely fine aluminium powder and tiny plastic beads that help the powder to flow evenly. The knobs on either side are connected to horizontal and vertical rods by very thin steel wire. The stylus is controlled using this pulley system, which in turn scratches off the aluminium powder coating the screen and thus creates a black line. Shaking the Etch A Sketch remixes the powder and coats the full screen, rendering it a fresh canvas. Alternatively some people who are particularly proud of their drawings drain the rest of the powder from the device, keeping the drawing intact.

The Etch A Sketch demands a great deal of skill to create really elaborate drawings upon its screen, the most challenging test of a user's coordination being the creation of curved lines by turning both knobs at the same time. Furthermore, the line made by the stylus has to remain unbroken throughout, meaning that all accomplished drawings require a degree of forward planning.

1816
Sir David Brewster

KALEIDOSCOPE

Scottish scientist Sir David Brewster invented the kaleidoscope in 1816. Brewster devised the name "kaleidoscope" by amalgamating three Greek words—"kalos" meaning "beautiful", "eidos" meaning "form", and "scopos" meaning "watcher". The kaleidoscope in its early form was a tube containing loose shards of coloured glass, which were viewed via their reflections on angled glass lenses and mirrors, which replicated them to create appealing patterns at the end of the tube. This effect of reflective symmetry had been observed for centuries in one form or another. It is claimed that the early Egyptians were accustomed to placing slabs of highly polished limestone at various angles in order to view fascinating reflections and patterns created by illuminated dancers.

In the 1870s Charles Bush made improvements in the design of the kaleidoscope. Most noteworthy was the mix of liquid-filled glass, which contained air bubbles, that continued to move after the kaleidoscope was at rest; and the solid glass pieces of spectacular and well-chosen colours, which created beautiful patterns for the benefit of the user. This was called Bush's "parlor" kaleidoscope, and it was this design that was mass-manufactured.

Left A kaleidoscope manufactured by Charles Bush—the first major 'scope' manufacturer in the US—circa 1875.

1848
John B Curtis

CHEWING GUM

Throughout history people have chewed natural latex products—such as sap and resin from trees, sweet grasses and grains—whether for medicinal reasons or merely as a way of freshening breath. Both the Ancient Greeks and Mayans were known to habitually chew on resin. Traditionally made from chicle—a natural resin found throughout Central America—many modern chewing gum manufacturers now use rubber as part of their water-insoluble gum base, which is then combined with water-soluble flavourings, sweeteners and colourings. In 1848, American John B Curtis developed and sold the first commercial chewing gum under the name "State of Maine Pure Spruce Gum".

Chewing gum is believed to have certain beneficial qualities: ever since the First World War the United States Military has regularly supplied soldiers with gum as it has been proven to help soldiers' concentration and reduce stress. In fact, it was such an important element of wartime strategy that once the enemy discovered the tactic during Second World War, shipments of gum from the Gulf of Mexico had to be escorted by US submarines. Recent studies show that chewing gum can also improve one's mood. A downside being, however, that with some 935 million packs sold every year in the UK, the government spends an average of £150 million pounds every 12 months cleaning the discarded gum off the streets—a clean-up cost which has prompted Singapore to ban the sale of the product in the city altogether.

25¢
QUARTERS ONLY

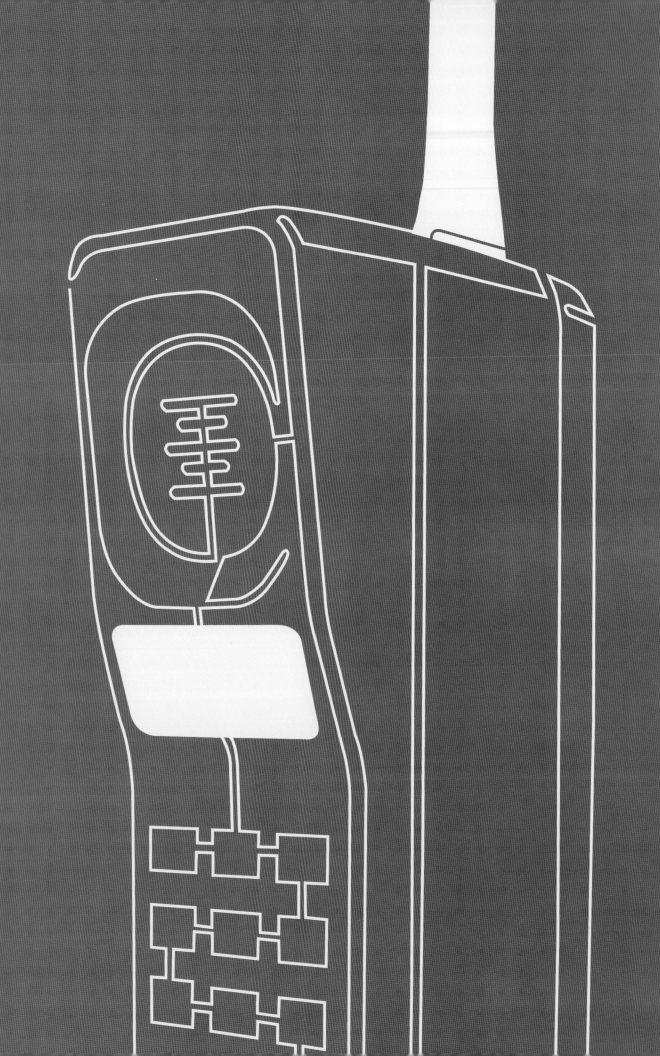

communication

PENCIL

The first modern-style pencils were produced with the discovery of the first graphite mine, at Seathwaite Fell in Cumbria, England in 1564. Graphite had virtually the same properties as lead; when newly-discovered it was called "plumbago" from the Latin word for lead ore, the major differing factor being that pure graphite is composed of carbon. At this time the English monopolised the production of pencils as no other graphite mines were known. The method of production involved sawing the graphite into sheets and then into square rods, which were inserted into hand-carved wooden cases.

The major breakthrough in pencil technology is attributed to the French chemist Nicolas-Jacques Conté, who developed and patented a process to make pencils in 1795. Although the now famed Faber family of Germany was producing pencils prior to this, they were crude and impractical, being made from ground graphite which was prone to breaking. They soon took to Conté's method, which involved drying and grinding graphite and then mixing it with clay. This doughy mixture was then pressed into sticks and kiln-fired. These rods were then inserted into a wooden case, usually made from pine or redwood. The most ingenious part of the process was that, by varying the ratio of graphite to clay, a pencil can be made to a specific hardness, subsequently differing the way the contents appear on paper. This process and basic recipe is still used, almost without differentiation, by manufacturers today.

Graphite-based pencils can now be found in around 19 different degrees of density and intensity and in myriad colours. Certain designs can now be used to write on surfaces such as plastics, cloth, film, wood, amongst others, as well as being water-proof and nonfading.

1847
Therry des Estwaux

PENCIL SHARPENER

French mathematician Bernard Lassimone applied for the first patent on a device—other than knife—to sharpen pencils in 1828. However, it was in 1847 that a fellow Frenchman Therry des Estwaux invented the manual, hand-held pencil sharpener that represented the device as it exists today in its most basic form. After this, the Love Pencil Sharpener was designed and produced in 1897 by America inventor John Lee Love, and was originally meant for use by artists. It was a portable, desk pencil sharpener, where the pencil was placed in a hole in the device and then a handle was rotated manually to sharpen the end of the pencil. Love's design was the catalyst for the expansion of the pencil sharpener industry and the design and development of the mechanical and electric sharpener.

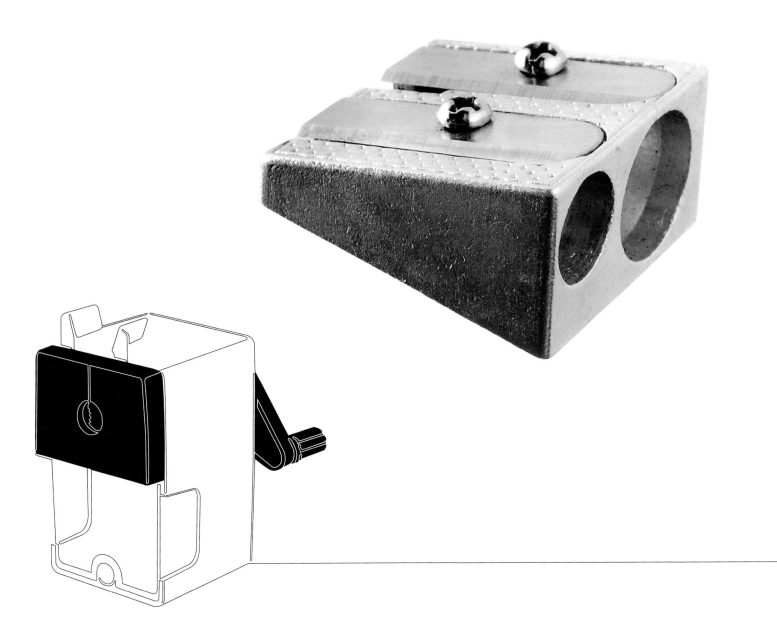

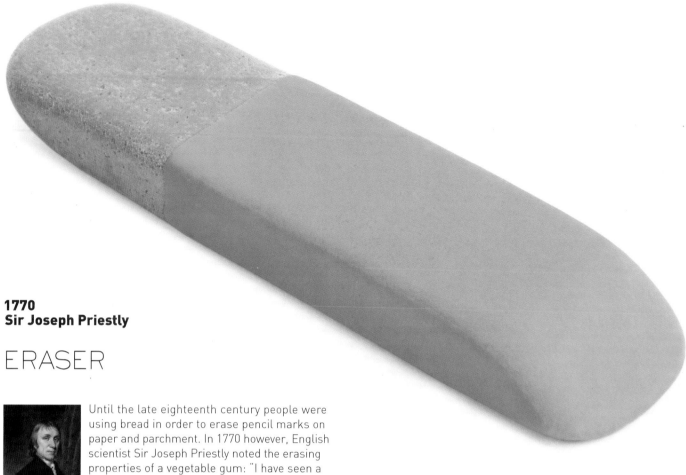

**1770
Sir Joseph Priestly**

ERASER

Until the late eighteenth century people were using bread in order to erase pencil marks on paper and parchment. In 1770 however, English scientist Sir Joseph Priestly noted the erasing properties of a vegetable gum: "I have seen a substance excellently adapted to the purpose of wiping from paper the mark of black pencil lead." He dubbed the substance "rubber". In the same year, an English engineer by the name of Edward Nairne inadvertently picked up a piece of rubber instead of a piece of bread and discovered its erasing properties; the natural eraser became widely marketed and was sold in half-inch cubes.

But natural rubber, like bread, was perishable and rotted quickly rendering it unusable. It was not until 1839 that Charles Goodyear developed and patented a process by which to cure the natural rubber and consequently instil durability within its properties. This process was known as vulcanisation—after Vulcan, the Roman God of fire—and involves cooking the rubber under pressure in order to cure it.

Modern-day erasers are either made from synthetic rubber or vinyl; both go through much the same process but only synthetic rubber erasers need to be vulcanised and tumbled. The raw material is mixed to its proper consistency and then put through a machine called an extruder which forces the material through a small hole, producing a long strand which is cut to three-foot lengths and then chopped into smaller 'plugs' by a rotary cutter. These plugs are then placed in a tumbler to round off the edges.

Erasers work on the basic premise that the molecules in an eraser are effectively 'stickier' than that of the molecules of paper, meaning that graphite pencil marks will stick to the eraser rather than the paper. Some erasers, such as those attached to the top of pencils, can remove the top layer of paper and leave a residue. In fact, American Hyman Lipman patented the rubber-topped pencil in 1858, but it was later invalidated due to its status merely as an amalgamation of two previously existing products.

700

QUILL PEN

The quill pen was introduced around 700 AD, remaining largely unchanged and unmodified until the introduction of the first prototype fountain pens in the eighteenth century. Its thousand-year dominance is testament to the effective simplicity of its design and ease of production.

The quill pen is made from a single bird feather, with different breeds affording varying characteristics to the instrument. Swan feathers were of premium grade and the most expensive due to their scarcity, but crow feathers were regarded as the best, and were used for making fine lines on parchment. Goose feathers were the mostly widely used because of their ready availability as a consequence of geese being kept in high numbers for agriculture. Goose feathers contained a unique feature that made them very practical for writing: the ability to hold and then release a flow of ink when pressure was applied to the shaft. Other less commonly used feathers were that of the owl, hawk, turkey and eagle.

It was always beneficial to take the feather from a living bird during the spring as these made the strongest quills. Furthermore, flight-feathers of the left wing were preferred as the feathers curled outwards when held by a right-handed writer. These feathers were then cut down to a length of between six and eight inches, thoroughly cleaned and scraped, and then hardened by being left in the sun to dry for around a week. After this, the nib was cut with a traditional pen-knife—a very sharp fixed-blade knife solely for the cutting of nibs for quills, unlike the miniature pen-knives of the present day—a long preparation process for a writing instrument that only lasted between a week and two weeks of use.

In actual fact, the feathers on the quill pen were not necessary and as early as the mid-seventeenth century most of the feathers were removed from the quill if not all of them; it is in this latter stage of the quill's development that it begins to resemble the modern day fountain pen.

FOUNTAIN PEN

The basic modern fountain pen contains a refillable reservoir of water-based ink that is drawn to the paper through a nib by a combination of gravity and capillary action. The earliest record of a writing device using such a principle dates back to tenth century Egypt, but the design failed to advance much until the 1800s due to little understanding of the effects of air pressure and the corrosive nature of the inks previously used. It was the invention of three key components that made the fountain pen a popular and widely used writing instrument: hard rubber, the iridium-tipped gold nib, and free-flowing ink. The 1880s saw the mass-production of the fountain pen, fuelled by insurance salesman Lewis Waterman who patented his design in 1884. Prompted by a leaking pen destroying an important contract and costing a business deal, Waterman took the initiative to apply the rules of capillary action to the instrument to reduce leaks. His product, christened "The Regular", was initially handmade and sold from a tobacco shop but, by 1889, Waterman had set up his own factory to produce a number of styles. Soon after his death in 1901, sales of his product, now marketed by his nephew, reached over a quarter of a million units per annum.

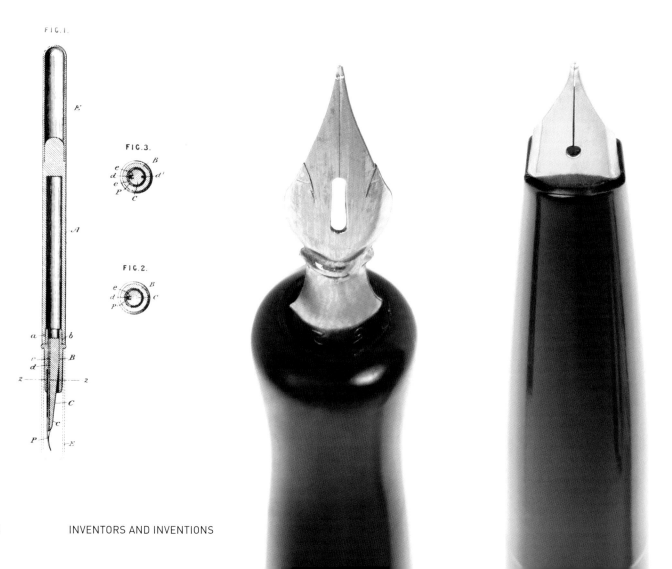

1931
László Bíró

BALLPOINT PEN

Frustrated by the failings of other pens and tired of refilling ink and cleaning up smudges, Hungarian journalist László Bíró was inspired to design his own writing device. With the help of his brother Georg, he set about inventing a pen that could use the same type of ink used in print—ink that dried much quicker but would clog up a conventional fountain pen.

The design they came up with involved a tiny ball bearing that was free to turn in a socket at the tip, lubricated by the ink, which it picked up as it rotated and then left on the paper. They presented the first version of their design at the Budapest International Fair in 1931 before patenting it in 1938. Two years later they fled the Nazi regime to live in Argentina, where the first commercial version of their pen appeared in 1945 through Eversharp Co and Eberhard-Faber. They next planned to approach the US market but Chicago businessman Milton Reynolds got there before them, ignoring patent rights and launching his Reynolds Rocket pen around the same time.

The ballpoint pen really took off when the patent rights were obtained by Frenchman Marcel Bich, who used a new manufacturing process and thick paste-like ink to create a reliable and inexpensive writing tool under the name Bic, in 1950. Around 14 million of these disposable pens are sold every day and, in September 2005, Bic sold their 100 billionth, making it the bestselling pen in the world.

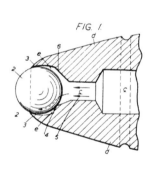

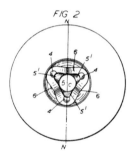

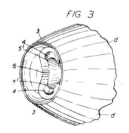

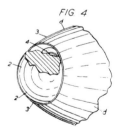

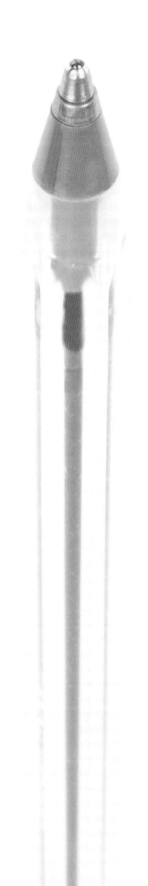

GUTENBERG PRESS

It has been estimated that the printing of literature was developed as early as 700 AD in the East, although the earliest dated book is *The Diamond Sutra*, printed in China in 868 AD. It was also in China that the first movable clay type was invented, but these printing methods never diffused through to Western civilisation, thus creating a geographical disparity and the belief that the subsequent first developers of Western presses would have been ignorant to the processes abundant in the Far East.

In 1439 a German, Johannes Gutenberg, invented a pressing process that revolutionised the world of printing, accelerated the spread of Renaissance culture, and consequently contributed to the evolution of Western society. Like most European printers, Gutenberg was using xylography (the art of wood engraving and bock printing), which involved a great deal of time and expense because it needed to be hand carved and the type was easily broken or eroded. Gutenberg realised that a new method was needed to supply a growing European demand for literature.

A former goldsmith, Gutenberg's solution developed from his professional experience. He produced a metal alloy that could be melted at low temperatures and poured it into a mould to create individual, hard-wearing, metal type that cast well in dye and could be used repeatedly and produced quickly. Ingeniously, the individual type was movable and could be arranged in any alignment desired. Individual letters were put together to form words; words separated by blank spaces formed sentences, and then these lines of type were brought together to make up a page.

One problem encountered was how to replicate identically the intricate calligraphy found in the books of the fifteenth century. This was overcome by using the skills of a calligraphic artist when casting the types. There now derived two sets of type for every letter: the standard separate form and the linked ornate form; once these letters were typeset they were pressed on to the page using a press that was based on already existing models used for making cheese, wine, and paper, with an operator working a lever to increase and decrease the pressure of the block against the paper.

An immediate effect of the printing press was that it made information available to a wider population of society and began a communication revolution, as well as a pioneering step in the provision and sharing of information. This enabled the spread of new ideas quickly and with great impact, comparable to that of the internet today.

The Gutenberg bible—one of the first printed books in Europe—printed by Gutenberg in Germany during the 1450s.

TYPEWRITER

The concept of the typewriter dates back to 1714 when an Englishman, Henry Mill, patented the idea of "an artificial machine or method for the impressing or transcribing of letters singly or progressively one after another". Its purpose revolved around overcoming the slow pace and frequent illegibility of handwriting. However, the vague idea stayed exactly that, and it remains unknown whether a prototype was actually produced. Over the next century and a half there were various unsuccessful attempts at constructing a practical typewriter, but all were slower than actual handwriting; they were inefficient, and some were as big as pianos.

Around 1868, Christopher Sholes, a journalist for *The Milwaukee News*, alongside his associate Carlos Glidden, patented a new typewriter much faster than a pen that was developed from an idea described in *Scientific American* by Englishman John Pratt. The printing type was mounted on the end of a type bar, and when a key was pressed it swung the type bar up to the paper. This is known as an upward striking typewriter, or "blind writer", as you could not check your work whilst writing as it was hidden from view. The secret of its success was, paradoxically, that it slowed the typist down. The inventions investor James Densmore devised a keyboard layout that is still used today, known as the Qwerty sequence; by placing frequently used letters further apart the typist would have to move their fingers far apart, slowing down the speed at which the keys dropped and therefore eradicating the problem of jamming.

The original Sholes & Glidden typewriter, produced by Remington between 1874 and 1878, sold only 5,000 in five years; as a consequence Sholes sold his share to Densmore as he lacked the patience to market the invention. However in 1878 the Remington No. 2 machine was introduced with the option of lower and upper case using a shift key. This model became the first commercially successful typewriter over the coming decade and was the catalyst for a typewriter industry. By 1891, 100,000 Remington typewriters had been sold. By the first decade of the twentieth century the office typewriter evolved from the traditional 'upstrike' method to the, visible 'front-strike' typewriter, increasing efficiency and speed, and making the typewriter an integral part of office work and industry.

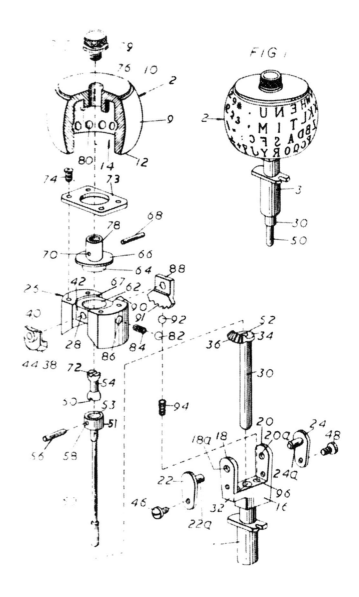

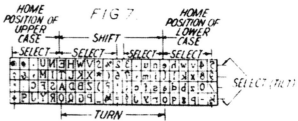

Top Illustrations of the IBM 72's 1961 patent golf ball head style typewriter mechanic, which allows the operator to produce work in a variety of different typefaces and change sector more easily.

Left An original Sholes & Glidden typewriter, circa 1874–1878.

Opposite top An Underwood Standard Portable Typewriter, encompassing a double shift feature allowing three characters per key.

PAPER

From the simple carvings and paintings on rock walls, to the incredible variety of paper and cards on offer in modern society, the medium of tangible visual communication, and the material on which man records such messages, has evolved for thousands upon thousands of years. In 4000 BC, the Sumerians would carve messages into heavy clay tablets, to be cumbersomely transported to the receiver. As time passed different materials were tested to find the most suitable writing material: ceramics, cloth, silk, bamboo, and vellum were all used, but the most common was papyrus. Papyrus was a plant found by the River Nile that the Egyptians harvested, peeled, and sliced into strips that were then layered, pounded and smoothed to create a rudimentary sheet of paper.

It was the Chinese who created paper proper in AD 105. The eunuch Ts'ai Lin is thought to have been the first to create the material using the bark of a mulberry tree combined with bamboo fibres, hemp and flax. The crushed pulp mixture was poured onto a sheet of cloth, pulled through two woven split bamboo strips and then left to dry so the water would drain through the cloth.

By the fifteenth century, paper was being mass-produced in the Western hemisphere in watermills, using the beaten pulp of cloth rags. But an increasing demand for literature meant that, by the seventeenth century, most mills could not find enough materials to meet the overwhelming call for printing supplies. By 1840 papermakers were using the cheaper and more plentiful resources of wood to produce sufficient amounts, though the process was still a slow and laborious one; this ended with the invention of the mechanical papermaking machine by Frenchman Nicholas Louis Robert in 1798.

1884
Ottmar Mergenthaler

LINOTYPE-COMPOSING MACHINE

In 1884, the German inventor Ottmar Mergenthaler produced the Linotype-composing machine, an invention so impressive that Thomas Edison proclaimed it the "eighth wonder of the world"; it was regarded as the most pioneering development in the printing industry since the invention of the movable-type Gutenberg printing press some four centuries previously. Gutenberg's printing process of typesetting individual letters by hand was now outdated, inefficient, and too slow for the fast-moving modern Europe. The Linotype machine allowed whole lines of type to be set at once using its 90-character keyboard; these solid lines of type were colloquially termed as "slugs".

The Linotype-composing machine was seven feet tall, seven feet deep, and six feet wide. It was operated by one person and was four or five times faster than traditional printing processes, enabling an operator to produce between five and seven lines of type per minute. In effect the operator was machinist, type-setter, justifier, type-founder and type-distributor amalgamated into one, causing many hand compositors to lose their jobs.

The composition process involved entering the text for a line on the keyboard, complete with spaces and punctuation, as well as letter matrices. These matrices travelled through channels where they were then aligned in the 'assembler' in the order they were released. Once no more text could be fitted on its individual line the 'casting' lever was pulled, which lifted the completed line from assembler to the casting section. The remainder of the process is automatic and the operator could then proceed with assembling another line of text. These 'slugs' of text were then arranged accordingly and pressed on to paper to create a full printed page.

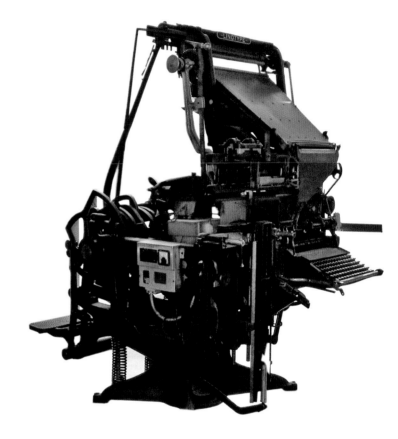

Top A *Le Monde* linotype machine.

Bottom Linotype machines in use at a New York printing room, 1909.

1843
Alexander Bain

FAX MACHINE

Faxing—the fax drawn from the word facsimile— involves the transmission of data, drawing or writing, and provides the receiver with a hard copy of the transmitted data . It was Alexander Bain, a Scottish clockmaker, who first received a British patent for the precursor of this telecommunication technology in 1843. He defined his developments as "improvements in producing and regulating electric currents and improvements in timepieces and in electric printing and signal telegraphs". To send a message via Bain's device an electrically conductive material would be used to write something or draw a diagram on a piece of paper, which then would be wrapped around a rotating drum situated under a needle attached to an electric pendulum. As the needle passed over the writing or drawing, electrical impulses were generated and transmitted to a second pendulum, connected by a wire and swinging in synchronicity with the reader pendulum, causing it to replicate the marks on another piece of paper.

Although Bain never built an actual fax machine it is clear that his invention made facsimile transmission feasible; many commercial machines were based on his design, one of the first of which was introduced by an Italian abbot in 1865. Giovanni Caselli's "pantelegraph" was used by the French Post & Telegraph agency to send messages between Paris and Marseilles, and later between London and Manchester. The problem was that, even though the fax machine utilised far superior technology, people could make do with the telephone and telegraph, which had a monopoly on the telecommunications market. The pantelegraph was seen as a means only for sending drawings, an outlook that directly appealed to the Emperor of China; because of the intricacies of Chinese text with thousands of ideograms, it was nigh on impossible to take advantage of telegraphy, but the pantelegraph would be able to accommodate these intricacies. Nonetheless, talks between Caselli and Peking fell through, and Caselli never lived to see his invention make a significant impression in the world of communications.

Despite being over 155 years old, Bain's basic design still remains the basis for modern fax machines, although now the images or words are digitised into a grid of dots and broken down into a binary code: white dots are equivalent to 0 and black dots equivalent to 1, creating what is referred to as a "bitmap", which is transmitted and received as data by the remote fax machine and reproduced as a hard copy. This process of photoelectric scanning and transmission was invented by German physicist Dr Arthur Korn in 1902; and by 1910 the system had become a must-have for the newspaper industry. During the 1920s the fax machine was developed for use as a domestic appliance, but it was not until the 1960s that the fax left the newspaper offices and began to make an impact on the business world as a whole. The turning point came in the 1970s when Japanese companies began making fax machines that were smaller, easier to operate, more efficient, and importantly, cheaper to buy.

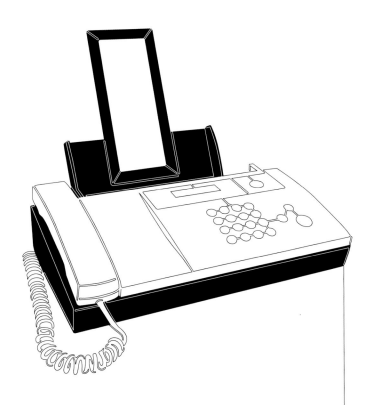

1822
Charles Babbage

COMPUTERS

Top Apple iMac computer.

Bottom The ZX Spectrum is an 8-bit personal home computer released in 1982 by Sinclair Research Ltd.

Opposite top right A vintage computer.

Opposite bottom Patent for an Electronic Numerical Integrator and Computer, filed 26 June, 1947.

Early systems of calculation involved vast tables of logarithms, overseen by mathematicians who were known as 'computers' but often compiled by young workers who each had a particular part of the formula to work out. These disparate parts of the equation would then be compiled to give a final figure. The fractured nature of this process meant that the outcomes of such substantial mathematics were often inaccurate and riddled with mistakes.

Englishman Charles Babbage realised this and was determined to eradicate these errors of calculation by simply removing humans from the equation altogether. In 1812, Babbage, a 21 year old Maths student at Cambridge, had the idea of designing a machine that could calculate these tables automatically. Ten years later he had built the first working prototype of what he termed the "difference engine". Through a mechanical system of gears and axles it calculated complex mathematical equations through relatively simple component parts. Once an equation was completed the engine would print the figure, leaving no need for human intervention in the process. The machine was hugely sophisticated, but Babbage's government funding was cut as he consistently heightened his monetary requests whilst making relatively slow progress on his designs. A decade later, he designed the so-called "analytical engine", which would become the forerunner for modern computers. Conceptually, its complexity and sophistication was incredible; it would actually solve equations with components that are now part of every computer—such as a mechanical memory and central processor—and it could be programmed using punch cards. Unfortunately, an example of the machine was never made in Babbage's lifetime, though he worked on its designs until his death in 1871.

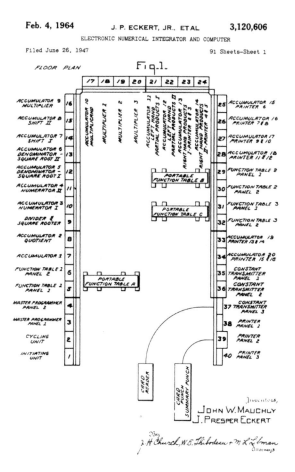

The next significant progressive step in computing came with German pioneer Konrad Zuse, building his Zi computer in 1931. It successfully performed calculations by the same methods as modern computers, using binary code signified by the use of electromechanical relays. However, similarly to Babbage, the German government paid little attention to Zuse's ideas and machines. Across the Atlantic, John Vincent Atanasoff designed the ABC computer in 1939, which used vacuum tubes as well as binary code to carry out calculations; he showed his unpatented designs to John Mauchly and John Eckert at the University of Pennsylvania, who subsequently took three years finishing their own ENIAC machine in 1946. The machine weighed 30.5 tonnes and contained 18,000 vacuum tubes, of which one would break on average every seven minutes when performing calculations. The computer also contained only 16K of memory, and had to be completely re-wired every time it was programmed.

Harvard University's Mark 1 machine, devised by Howard Aiken and built by IBM in 1944, was the first to have a series of stored instructions on paper tape. It was from this machine that references to 'bugs' in modern-day computers were derived; when once opened for repair, it was found that a moth trapped inside had caused the breakdown.

British Post Office employee Thomas Flower invented the machine regarded as the first computer that shares a resemblance to today's models in 1943. The "Colossus" was built to decode German military messages. Its unique purpose meant that the technology behind it was kept a government secret as far past the Second World War as 1976. The Univac I commercial model computer had been launched in 1950, though only 46 were ever built. It was not until the 1960s, with the invention of the electronic computer that the accuracy and technology of these early models was finally superseded.

1969
Advanced Research Projects Agency (ARPA)

INTERNET

In 1969, the Advanced Research Projects Agency—or ARPA—which funded research teams involved in projects of military significance in the United States, developed a computer network system called ARPANET. ARPANET, in effect, worked as a trial run of the internet we use now. These researchers had access to only four computers but they were spread widely across America, in Los Angeles, Santa Barbara, Stanford, and at the University of Utah. All were hooked up through 56 kilobit-per-second links to ARPANET to allow any one of the researchers to access any one of the computers at any time. The network system continued linking up computers in research labs throughout the 1970s.

To make the network operate and allow different computers to be able to communicate with each other, technology such as Internet Protocol and the File Transfer Protocol had to be developed; these relied heavily on what is known as 'packet switching', a cornerstone of the internet's operations today. Through packet switching a document, image or sound is passed along a network. To do this, the information is broken down into small electronic envelopes of identical sizes, each with their own ID number; they are reassembled at their destination, all within a matter of seconds. Ray Tomlinson of ARPA pioneered email based on this system in 1971. Although messages could be left for other people in an inbox if they shared a computer, the real feat was to try and send messages to different computers. Tomlinson invented a protocol to achieve this and sent his first message to a computer less than a foot away from his; it read "QWERTYUIOP", the result of him dragging his finger across the top of the keyboard. Tomlinson admitted that his new protocol did not have a huge effect upon its arrival but that it was just a "neat idea"; however, its eventual marriage to the Internet turned email into one of the most important computer applications of all time.

These early experiments into the delivery of digitised information from computer to computer set the foundation for what would then proliferate across the world. What began as a small project involving only the scientists at CERN, a European research centre near Geneva, developed rapidly into the 'global read-write information space' that Tim Berners-Lee proposed in 1989—now commonly known as the World Wide Web, or internet.

1901
Reginald Fessenden

RADIO BROADCASTING

The history of radio broadcasting, and the technology that afforded it, developed from the use of spark-gap telegraphy. The aim was to develop the capacity to transmit the nuances of the human voice—a full range of sound, rather than just the dots and dashes of Morse code. A pioneer in this field was Reginald Fessenden, who was previously Thomas Edison's chief chemist and worked on improving the carbon microphone receivers used in telephones and developing audio reception of signals.

In December 1900, he was successful in transmitting a voice signal over 15 metres using a high-frequency spark-transmitter. This was achieved by combining two frequencies to form a third audible tone. He was awarded a patent in 1901 for a high frequency, continuous-wave voice alternator-transmitter, which was the early predecessor to the AM (amplitude modulation) radio system. The developed model was given an extensive demonstration to telephone and telegraph companies at Brant Rock, Massachusetts, on 21 December 1906. A few days later, Fessenden's device made history by delivering the first entertainment broadcast on radio, if only to wireless operators on board United Fruit Company boats in the North Atlantic. It comprised of Christmas carol being sung by an unknown woman, a violin rendition of "O Holy Night" played by Fessenden himself, and a passage from the Bible.

The next major development came in the form of amplification. In 1906, an American scientist, Lee De Forest, invented the Audion tube, a three-electrode vacuum tube capable of electronic gain by using a weak electrical current to control a larger one; in essence it was possible to detect and amplify weaker signals than had previously been possible. This allowed for long-wave transmissions by a multitude of radio stations previously unattainable with spark transmitters, giving birth to the invention of amplitude-modulated or AM radio.

De Forest's Audion tube paved the way for pioneering work in audio transmission, most notably the work of Edwin Armstrong who, in 1914, developed and patented the "regenerative circuit", a "wireless receiving system" that increased the sensitivity of receivers several thousand fold. The development of the device was claimed by numerous other inventors, including De Forest, but as these patent battles were carried out, Armstrong went on to invent the "Superheterodyne circuit" in 1917, and eventually, by 1933, received a patent for his creation of what we now know as FM radio: wide-band frequency-modulation radio. This involved varying the frequency of the wave rather than the amplitude, which enabled exact tuning to individual frequencies and utilised the whole frequency spectrum, resulting in reduced static and improved sound clarity.

There are many claims as to which was the first radio-station to start broadcasting for entertainment purposes, but one of the first was a radio station called 2XG, an experimental station set up by De Forest which started broadcasting phonograph records in the spring of 1916. It was not until the 1920s that commercial radio stations started broadcasting at a mass level in both Britain and the USA.

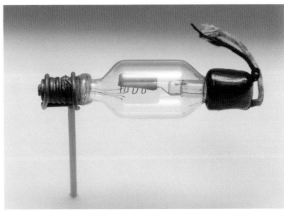

Top to bottom Radio valves; audion triode tube 1906; Brant Rock radio tower 1910.

1877
Thomas Edison

PHONOGRAPH

The idea for the phonograph first surfaced when Thomas Edison was working to increase the efficiency of a telegraph transmitter. He noticed that when played at high speed the noise produced resembled that of a human voice, and began to realise that it might be possible to record a telephone message. Thus, he began trialling a device that made use of the diaphragm of a telephone receiver with a needle attached to it. The premise of the invention was based upon the fact that the needle would move up and down in accordance with the vibrations of the voice, making indents into paper tape and recording what had been said. Edison's experimentation lead to the creation of a tin-foil cylinder version, indented with a stylus and utilising another needle for the playback of the recording. The first message ever recorded and heard back was Edison reciting "Mary had a little lamb." In 1877, the year of the phonograph's first prototype, the inventor presented his device to the staff of Scientific American in New York; the 22 December issue reported that "Mr Thomas Edison recently came into this office, placed a little machine on our desk, turned a crank, and the machine inquired as to our health, asked how we liked the phonograph, informed us that it was very well, and bid us a cordial good night." It was the first machine of its kind and brought Edison worldwide fame.

In 1878, The Edison Speaking Phonograph Company was founded and the instrument was put into production. Although the invention was an instant success, it was acclaimed only as a novelty item, as a consequence of its impracticality: it was difficult to operate and the tin-foil cylinder deteriorated after just a few

uses. Its popularity slumped as the novelty value dwindled, causing Edison to stop development and concentrate on the invention of the incandescent light bulb instead. Despite this, he did offer some practical uses for the phonograph in the future: dictation in offices, phonographic books for blind people, the teaching of elocution, and the playing of music were all cited as possibilities.

It was during this period that other people began to develop the phonograph further, most notably Alexander Graham Bell. He and his partner, scientist and instrument maker Charles Sumner Tainter, made improvements to Edison's invention; they made the stylus floating rather than rigid because of the problem with piercing rather than indenting cylinders, and also changed the material of the cylinder from tin-foil to wax. They were awarded a patent on 4 May 1886 for their improved phonograph to which they had given the name Graphophone. Edison refused the offer of a possible collaboration from Bell and Tainter, wanting to personally improve his device, and in May 1888 introduced the improved Phonograph and the Perfected Phonograph which both made use of wax cylinders, made out of ceresin, beeswax, and stearic wax.

Jesse Lippincott was the businessman who brought both the Graphophone and Phonograph together, but only by becoming the sole licensee of the American Graphophone Company and then proceeding to buy the Edison Phonograph Company and many other smaller phonograph companies to form a conglomerate, the North American Phonograph Company. Lippincott saw the potential in, and focused on, the use of the phonograph as an office tool in the field of business. This company did not prove profitable and met with stern opposition from stenographers and, after Lippincott fell ill in 1890, Edison seized control of the company as its chief creditor before buying the rights to his invention back after declaring the North American Phonograph Company bankrupt in 1894.

In 1896, Edison started the National Phonograph Company, manufacturing phonographs for home entertainment, with models

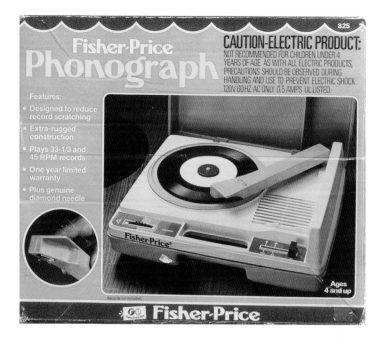

such as the Spring Motor Phonograph, the Edison Home Phonograph, and the Edison Standard Phonograph, as well as the commercial sale of cylinders. The standard size of a cylinder was 108 mm long and 555.6 mm in diameter, played at 120 rpm, but only played for a two-minute length of time. They were also costly as there was no way, as yet, to produce cylinders at a larger scale, so every recording was individually made every time. In 1901, a process for mass-producing wax cylinders was devised, whereby the cylinders were moulded from a template rather than indented by the stylus, and moulded in a harder, more durable wax. This process was called Gold Moulded in reference to the gold vapour that was produced when moulding.

Improvements made to the machine over the years meant that the models were still significantly more expensive than the similar Dictaphones sold by Columbia—Edison's main business rival—and the cylinders were more expensive, and with shorter play times, than competitors discs which offered up to four minutes. Many of Edison's main competitors, abandoned the cylinder market, but Edison stubbornly pursued the development of wax cylinders for the Phonograph eventually creating, in 1913, the Blue Amberol Record. It was an unbreakable cylinder with arguably the best sound quality at the time but, even with this superior quality, the rival disc was at the peak of its popularity, and Edison could not break the market. He continued to make his novel cylinders but yielded to the popularity of the disc by introducing the Edison Disc Phonograph in late 1913.

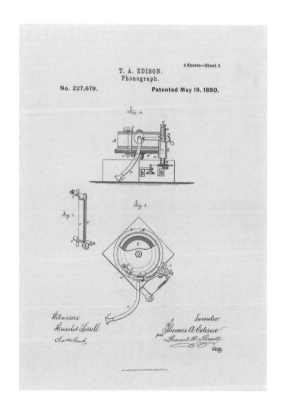

Opposite top An Edison cylinder phonograph, circa 1899.

Opposite bottom Thomas Edison and his phonograph.

THE POSTAL SERVICE

As early as 2000 BC the ancient Egyptians were using a rudimentary courier service where the written decrees of pharaohs were delivered around Egypt to inform his populace. But it was an astute and lateral-minded Englishman who devised the first genuinely efficient national postal system.

Rowland Hill was a Victorian administrator and tax reformer who, in 1837, released a pamphlet indicating the inherent flaws in the current postal system. His argument was that it was unnecessarily complex and expensive: the cost of every single letter had to be worked out according to its exact weight and exact destination. He demonstrated that the costs and time spent delivering a letter were not incurred in its transportation but rather in its handling and sorting, a time consuming process that Hill believed could be streamlined. Hill's proposal was that there should be a flat fee to be paid in advance of one penny per half-ounce, denoted by adhesive stamps, for delivery anywhere in Britain. The system was given full backing by the government and, on 1 May 1840, Britain saw the introduction of Hill's pre-paid adhesive stamps, the Penny Black and the Twopenny Blue. The system was an instant success in Britain and was rapidly adopted worldwide.

By the first quarter of the twentieth century, the sheer volume of letters being sent overloaded the network past the point of practical operation. Improvements to the system had to come quickly. One of the most innovative was a remote-controlled train: the Mail Rail travelled underground, carrying mail from the suburbs to a central London sorting office and cutting the delivery time from a 45-minute van journey to just six minutes. However, by the 1960s it was realised in most national postal services throughout the world that the way forward lay in mechanical sorting, which was facilitated by the use of postcodes or zip codes that were read using an electronic eye and then assorted into the appropriate bin, giving rise to fully automated post offices. The first was opened in Rhode Island, USA, in 1960. By the 1990s, machines could read handwriting using optical character readers; they could 'fast-track' 11 envelopes per second, 20 times faster than hand sorting.

Even with the advent of the internet and email the postal service remains a vital part of business and the infrastructure of any developed country, yet it is still viewed as a more traditionally personal way in which to communicate.

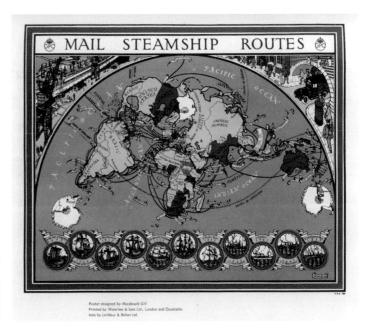

1876
Alexander Graham Bell

TELEPHONE

On the 14 February 1876, Alexander Graham Bell filed a patent for the telephone. He had devised a way to send speech through an electric current, improving and updating the telegraph system. Bell was not alone, however, other inventors were close behind—Elisha Gray, Philip Reis and, in 1877, Thomas Edison, who replaced the coil and magnet receiver with a button of carbon, which subsequently improved the sound quality. Edison's device evolved to include a hand crank, which signalled a call being made down the line, and a hand piece, which allowed the user to open the line. It took almost 20 years, however, for the telephone to catch on; in 1895 a reported 300,000 were in use. Over the next 50 years, the popularity of the device exploded, and by the 1950s nearly every house had a telephone. The next major development in telephone technology came in 1970s America, with the electronics company Motorola in New York developing a mobile phone the size of a brick. The phone used the technology that had been used for decades by the emergency services and the now archaic pager. Today, the mobile phone is available in a number of guises, all of which have an aim of being as convenient as possible. Moving from analog to digital, the mobile phone can now combine a number of functions: the internet, digital camera and MP3 player, amongst others.

Top right Telephone operators 1952, Seattle.

Opposite bottom left Patent for Nokia Mobile Phones Ltd filed 25th May 1989 by Jouko Tattari, Finland.

Opposite bottom right iPhone.

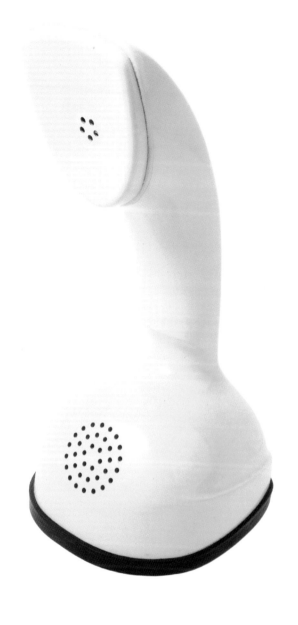

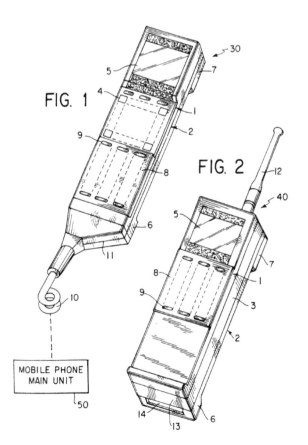

FIG. 1

FIG. 2

MOBILE PHONE MAIN UNIT

GLOBAL POSITIONING SYSTEM (GPS)

Ivan Getting, a research engineer at the Raytheon Corporation, who had a large hand in the development of wartime radar and microwave ovens, developed what is known as the Global Positioning System (GPS). Getting was asked by the American military to devise a system whereby they could fire guided missiles accurately while on the move so as to minimise the risk of attack. He proposed a network of radio transmitters each with an individual clock. The same type of device—along with a receiver—that could receive several transmissions simultaneously would be attached to the missile launcher. In a procedure not unlike that of Radar, the missile's minders would record how long each signal would take to be received from each one of the transmitters. This would allow the minders to calculate exactly where they were launching the missile from even while mobile. This meant they could attack their target accurately while preventing their target from accurately attacking them.

The US Military launched the first GPS satellites in 1974, although for over a decade only the military had access to the system. It was not until the 1990s that full civilian access was granted. Today's GPS involves a number of radio-transmitting satellites, each the size of a small car, circling the earth at precisely-known orbits 20,000 kilometres up. Three or four of these transmissions are used by the small GPS receivers to calculate its position in three dimensions: latitudinal, longitudinal and altitudinal. This provides the user with a very precise location, accurate to a few metres. Allegedly, the defense authorities now have sole access to an enhanced system that provides positions with even greater accuracy, working to the nearest centimetre.

These receiver units were adapted so that they could be used on ships and planes, and as a handheld model called "Scout" for hikers and orienteering. GPS technology has also since been incorporated into mobile phone technology. Perhaps more common is the use of GPS receivers in modern cars, which not only provide the location of the car but also plans a journey. Once programmed with a destination, it will even correct the mistakes made by the driver along the way.

1904
Christian Hülsmeyer

RADAR

Left Means of detecting aircraft by using radio waves, filed 17 September 1935 by Robert Alexander Watson-Watt for the National Physical Laboratory, Teddington, England.

Right Radar dishes.

First patented in Germany by Christian Hülsmeyer in 1904, the Telemobiloskop, or Remote Object Viewing Device, was initially used as a system to avoid ship collisions, but the technology fell from public view due to a lack of commercial interest. Radar, as the system would come to be known, found its true success as an aeronautical and military application, as a means of detecting aircraft via the use of radio waves. It was developed in this respect by Robert Watson-Watt, the Superintendent of the Radio Research Laboratory in Britain, in 1935, although TV pioneer John Logie Baird had previously described "a method of viewing an object by projecting upon it electromagnetic waves of short wavelength" seven years earlier.

When Watson-Watt was first contacted by the Committee for the Scientific Survey of Air Defence, their interest in radio waves was far more sinister; they were curious to know whether radio had 'death ray' potential—whether waves could be used to debilitate enemy pilots by heating their blood. Watson-Watt replied that this was not possible because of the amount of power necessary for such a result, but added that the detection of aircraft was a more feasible goal. By bouncing radio waves off aircraft in the same manner as light, and timing the gap between the 'echo' of a radio pulse sent by a transmitter and the moment at which it was picked up by a radio aerial, the distance and position of the aircraft could be calculated.

The Germans were already aware of this technology and had already begun research into its practical utilisation. The British, therefore, had to move fast in order to gain the upper hand on their enemy. Subsequently, they became the first to trial the technology on 26 February 1935. When the Heyford bomber flew only 2,000 metres away from a transmission station in Daventry, it was identified as being 13 kilometres away on a cathode-ray oscilloscope display.

After further tests proved more successful, systems were developed to calculate and measure distance, height, direction and numbers of attacking enemy. The first purpose-built station was erected on the Suffolk coast where it could detect aircraft up to 64 km away. This was the catalyst for the construction of another five stations around the Thames estuary of London and, by early 1939, another 20 had been placed around the country from the Isle of Wight to Dundee, capable of spotting enemy warplanes over 70 miles away. This was largely due to the invention of the cavity magnetron by John Henry Boot, which gave a compact and powerful source of short-wave radio waves for detecting craft at greater distance.

The technology first became known as RDF (Radio Direction Finding) but latterly gained the America-coined moniker RADAR (Radio Detection And Ranging) in 1943. The system was invaluable to the British during the Battle of Britain in 1940 as it allowed the capacity to detect where the bombers were intending to attack in advance, meaning that relatively small fleets of the RAF could be strategically placed to fight the oncoming bombers. It was an absolutely vital invention and was indispensable in winning the war. Not only was it able to reduce the damage inflicted during the night-time bombing of the Blitz, but it also had utmost importance in the victory of the Battle of the Atlantic by finding German U-boats, as well as guiding British bombing raids into occupied Europe. Watson-Watt was awarded a knighthood in 1942 for his achievements in the field.

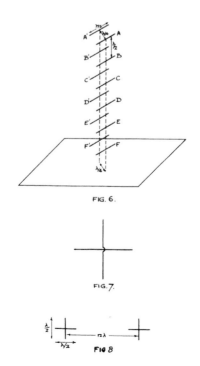

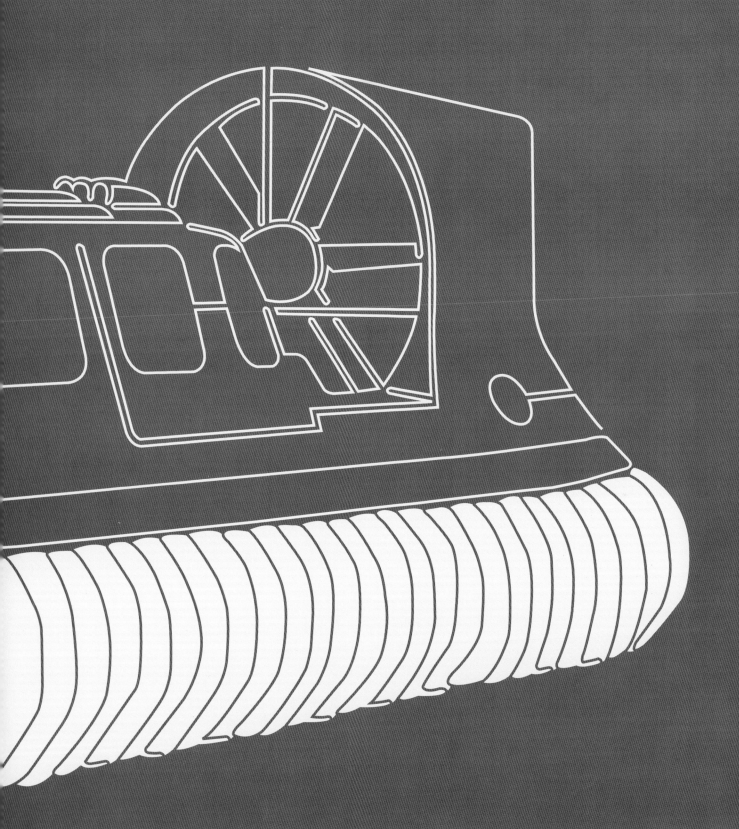

engineering
and
transport

ARCHIMEDES SCREW

A device whose structural simplicity belies its engineering logic, the Archimedes Screw is an object with a great history of agricultural, mechanical and logistical applications. Attributed to Archimedes in 300 BC, the screw in its original form conveniently transferred water from low-lying to raised areas and was initially used to transfer water to irrigation ditches.

Working on the Archimedean principle of buoyancy—that the buoyant force on a submerged object is equal to the weight of the fluid that is displaced by the object—the Screw essentially displaces water from one area to another using a force of downwards exertion.

The physical components of the Screw can be conceptualised by the image of an inclined plane enclosed in a hollow cylinder. More specifically, the Screw is formed by a metallic double or triple helix built around a heavy central pole that is enclosed—but not necessarily fixed—within a lightweight, cylindrical tube.

Traditionally powered manually, the Screw has since undergone rapid and inevitable modernisation as a result of industrial advancement, evolving into a mechanism that can be driven by both electronic and pneumatic motors. The lower end of the apparatus sits submerged in water and as the internal coil rotates, water is raised to the top via a method of transferal—as the bottom coil turns, water is caught in its planes and is scooped into the cylinder, moving upwards with the internal spiral until it passes out at the top. With water continually leaking from one chamber of the screw to another, a mechanical equilibrium is achieved, ensuring efficiency is kept to a maximum.

The Archimedes Screw has, since its invention in 300 BC, had its uses spread to a more diversified industrial field. Appropriating the concept of proportional transfer, the mechanics of the Archimedean Screw can be found in machines involved in drainage, injection molding, sewage purification and the handling of light and low density materials such as sand, grains, cereals and ash. The Archimedes Screw is also popularly found in processes engaged with controlled distribution—in rotary feeding and combine harvesting—and although the driving forces behind the Screw have changed, the base components remain almost identical to their 300 BC design, demonstrating that the most intractable engineering mechanisms need not always be the most complex.

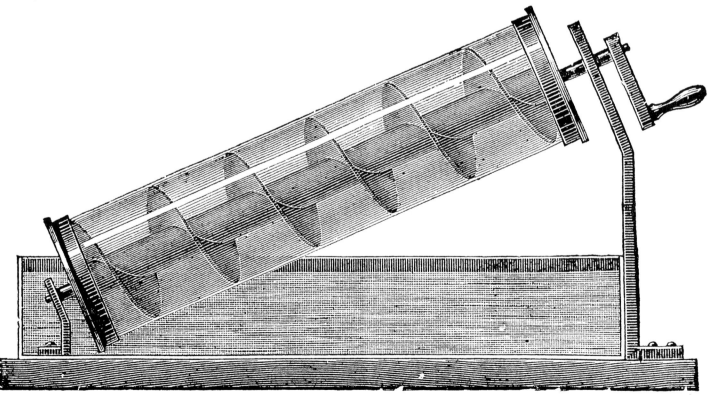

1794
Philip Vaughn

BALL BEARING

The ball bearing is one of the least acknowledged yet most important inventions of all time. Much of the industrial world as we presently know it is run upon the modest ball-bearing, an invention based on the ideas of Leonardo Da Vinci hundreds of years earlier.

In the sixteenth century, Da Vinci had an idea that described the mechanics of ball bearings and how they could be used in load-bearing movement. The purpose of the ball-bearing principle was to reduce the amount of rotational friction that occurs between the moving-parts of a bearing and to support radial and axial loads. The first patented design for a ball bearing was accredited to Philip Vaughn—a Welsh inventor and ironmonger—in Carmarthen, 1794. Whilst working on carriage axles he devised a system whereby the balls ran along a track in an axle assembly, thus creating the first ball-race system and the predecessor to modern day ball bearings. A 'race' is essentially a grooved track; a ball-race system features two such tracks, one fitting inside the other. Metal balls sit in the tracks and are contained by the races; the balls rotate with the races and transmit the load. The geometrical properties of a circle means that it has much less of its surface area in contact with its surroundings when rolling than two flat surfaces, and herein lies its success: less friction occurs, increasing performance.

Ball bearings are hugely important to the running of the industrial world, never better highlighted than in the Second World War. In 1947 the Allied Forces committed 300 USAF bombers to the destruction of a German ball-bearing factory in Schweinfurt, the idea being that without a supply of ball-bearings the wheels would, literally, fall off the Nazi campaign.

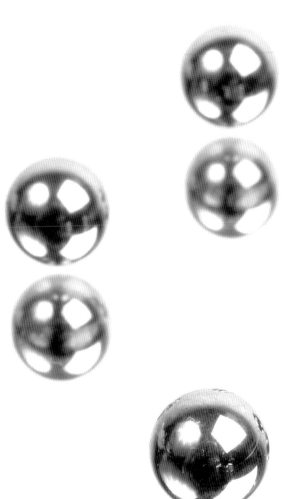

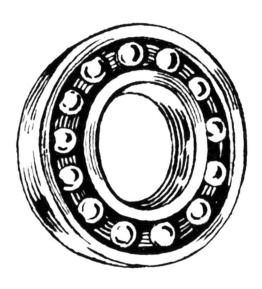

1545
Gerolamo Cardano

UNIVERSAL JOINT

The universal joint allows movement in any direction, along any axis, and also allows the transference of power. The device is essentially two rods connected at right angles to each other, conjoined by a cross-shaft using gimbals. It was Italian mathematician Gerolamo Cardano who first realised, in 1545, that the principles of the movement of gimbals could be applied to a device that would allow the transfer of rotary motion through an angled connection. However, like many of the great pioneers and figures of the Renaissance era, his ideas remained unrealised and no practical prototype was ever made.

Just over a century later the great English scientist Robert Hooke became the first individual to utilise the joint in an optical instrument, which was used to watch and track the sun. The handle, using the new joint, allowed the twisting movement of one shaft to be transferred to the other, in whatever position or angle each shaft was at.

It was in 1903, over two centuries after Cardano first came up with the joint, that a modified, sealed version designed by American Clarence Spicer was used for the transfer of power from the engine drive shaft to the axle of cars. The universal joint is now an automobile industry standard.

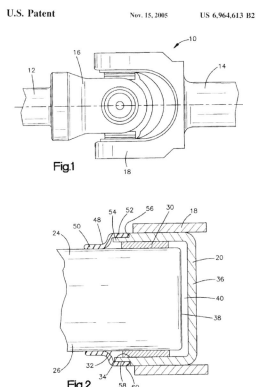

THE DYNAMO

U.S. Patent   Sep. 23, 1986   Sheet 1 of 2   **4,613,761**

FIGURE   I

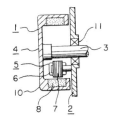

FIGURE   2

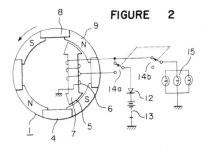

Hans Christian Oersted made the discovery of electromagnetism (the capability of electricity to create a magnet) in 1820, but it was Michael Faraday who set out to find whether the opposite could be achieved: could a magnet produce an electric current? Faraday's repeated experiments invariably failed from his first in 1822 up until 1831. He coiled 220 feet of wire around a cardboard tube and connected the two ends of the wire to an instrument that would gauge the electric current flowing through. He then used a rounded bar magnet eight and a half inches long and an inch in diameter to move in and out of the tube. When the magnet was pushed into the tube the needle would point in one direction and when taken out again, it would point in the opposite direction. When at rest the needle would not move, revealing that when the magnet was in motion an electric current would be produced via the breaking and connection of a magnetic current.

Faraday began designing a machine that would utilise his new discovery. A 12-inch copper disk was fastened on a brass axle, both of which had metal collectors attached to them, continuously spinning between the two ends of a mounted horseshoe magnet, which broke the currents of magnetism and produced a steady current of electricity. This was the first ever dynamo. After Faraday's invention of the very first electric transformer, many great names in the field of science, such as Edison and Tesla, set about improving the dynamo's design. The permanent magnet used by Faraday was eventually replaced by electromagnets, producing a much more powerful magnetic field. It could now be employed to power all sorts of operations and paved the way for some wonderful inventions, such as the modern electric motor, streetlights, traffic lights, electric-driven machines in factories and electric railways.

BICYCLE

The great Leonardo da Vinci sketched a design for a bicycle around 1490. The main purpose for designing such a contraption most probably grew out of a desire to replicate the movement of the treadmills of the Roman age, which made the transport of heavy loads quick and efficient by expending the least energy possible. Although never leaving the drawing board it shares a remarkable likeness to modern-day bicycles and was extremely innovative for its time; so much so that the first working model of anything resembling a modern bicycle did not appear until the late eighteenth century.

In 1797, French craftsman Comte Mede de Sivrac developed a wooden scooter-like machine called a "Celerifere". It was essentially a running machine, consisting of a wooden beam that had two wheels attached at either end, which the rider straddled, using their legs and feet to propel the device forward. It had no brakes or steering wheel, so long, straight, roman-style roads were necessary for a pleasant riding experience. 1817 saw further development in bicycle design via the German Baron Karl Von Drais de Sauerbrun; his model, named the "Draisine" or "laufmaschine"—meaning "running machine"—was entirely made from wood and had no pedals, meaning that no real mechanical connection was made between the rider and bicycle. It did, however, have a chest rest and, crucially, a steering wheel.

The first bicycle with pedals was invented by a Scottish blacksmith by the name of Kirkpatrick MacMillan in the 1830s. He never patented his invention as it failed to attract a local following despite his attempt at demonstrating its efficiency by cycling a 140 mile round trip from his hometown of Dumfries to Glasgow. However, it is the father-and-son team of French carriage makers, Pierre and Ernest Michaux, who are widely credited with the addition of the modern bicycle pedal and cranks, resulting in the French bicycle model known as "Velocipedes"—or, perhaps more accurately, "Bone-Shakers", the eponymous discomfort caused by the use of a huge front wheel, which not only increased speeds but also instability. The machines were mass-produced by the Michaux company between 1867 and 1870.

The Penny Farthing came after the French bone-shakers. Invented by British engineer James Starley in 1870, its design included a small rear wheel and a large front wheel, both of which were equipped with rubber tyres. By 1874 the design was improved using suspension for the wheels and wire spokes; it could reach a top speed of 20 mph, and was the first truly efficient bicycle. Unfortunately, and to many riders' peril, its tall-standing design made it treacherously dangerous, and so in 1885 the Starley Rover safety bike was born. Stability was achieved by returning the wheels to a reasonable size, and a chain-drive was added to the rear wheel to improve speed, as well as sprung saddles for comfort and a diamond-frame shape. Over a 20 year period the British engineers brought the bicycle to its basic present form until the addition of the derailleur gear system in the 1970s, which opened vast opportunities for bicycle design in the future.

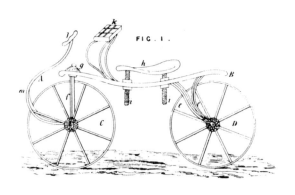

A.D.1818. DEC. 22. Nº 4321.
JOHNSON'S SPECIFICATION

FIG. 1.

FIG. 2.

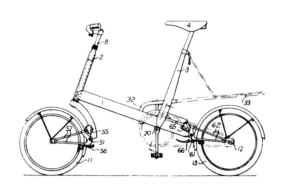

MOTORCYCLE

The first motorcycle was made in 1885 by a German, Gottlieb Daimler, and was called the "Einspur". Previous models were steam-powered but the design of the motorcycle was much more suited to the power-to-weight ratio provided by a petrol motor. Daimler's motorcycle was constructed mostly out of wood, with wooden spokes and iron-rimmed wheels; it had a four-stroke petrol engine attached to its frame and one stabiliser wheel either side of the rear wheel. Its 264cc of power provided the motorcycle with a top speed of 7.5 mph. Two Frenchmen, Count de Dion and Georges Bouton, were pioneers in the development of lightweight and powerful petrol engines that could be made in various sizes; because of these improvements, and a sheer disregard for patents, motorcycles began to be developed all over the world.

The first motorcycle to be put into mass production was another German model. The Hildebrand and Wolfmuller was a twin-cylinder, 1,488cc, 28 mph two-wheeler motorcycle patented in 1894 and built until 1897. Further contemporary developments in motorcycle design, increasing its usability and popularity, included the twist-grip throttle control applied to Americans' George Hendee and Oskar Hedstrom's 1904 Indian Single motorcycle.

The amount of motorcycle-makers rose rapidly at this time, and by 1914 there were 600 such manufacturers around the world. Motorcycles became extremely popular during the 1920s; in Britain alone there were 260,000 motorbikes in use compared to just 200,000 cars. It was not until the post-war baby boom that people were buying cars in larger numbers to cater for family transport and travel.

Today the motorcycle-manufacturing industry has been monopolised by Japanese racing manufacturers such as Suzuki, Honda and Yamaha, but the models that are most in demand are small and cheap, used in developing countries such as India and China. Motorcycles are used to great effect in built-up metropolises, given their mobility and capacity for squeezing through traffic where cars cannot. This laid the foundation for the development of a motorcycle 'taxi' industry, using mopeds and scooters to provide customers with a fast, if somewhat perilous, transport service.

From top to bottom Motorcycles, 1916; a Triumph engine in a Norton frame; the first petrol-powered motorcycle by Daimler featuring a wooden frame, solid tyres and small side-wheels for stability, 1885; the two hydraulic hoses entering the front fork of this Yamaha WR450F 2-Trac provide additional front-wheel drive.

2001
Dean Kamen

SEGWAY HT

American inventor and engineer-school dropout, Dean Kamen, unveiled the Segway Human Transporter in 2001 in the hope of revolutionising the way in which people use transportation, moving from A to B in their day-to-day lives. Kamen believes that his two-wheeled, self-balancing, electric-powered machine is capable of achieving this and much more by solving problems of pollution and congestion for an eco-conscious world. The designer has prophesised that his design will eventually "be to the car what the car was to the horse and buggy".

The concept stemmed from Kamen's witnessing the difficulty encountered by a disabled young man when he tried to mount a curb in his wheelchair; it was not due to a lack of effective design in relation to the wheelchair but the lack of balance that a disabled person has in a world built around able-bodied people. Kamen and his team set about restoring balance to people who could no longer walk, and created the Independent IBOT Mobility System based on self-balancing technology that allowed users to traverse rough terrain and obstacles such as stairs. An ingenious addition to IBOT was a facility that allowed users to elevate themselves to eye-level, providing them with a new perspective of the world. It was obvious that the balancing technology developed by Kamen and his team held exciting possibilities if it could be specialised for use by the able-bodied population, and it was this that led to the creation of the Segway HT.

The Segway HT uses unique technology called "LeanSteer" whereby the frame and handlebar sense the changing movements in body position and centre of gravity at about 100 times per second and react accordingly, balancing and stabilising using "five micro-machined angular rate sensors"; basically a system consisting of gyroscopes, tilt sensors and an inbuilt computer. So, a turning movement is controlled using the handlebars, but direction of travel, acceleration (to an optimum speed of 12.5 mph, and braking is controlled wholly by shifting the body's weight forward or backwards.

The design is electrically-powered and its batteries can be charged from a standard plug socket; charging it for somewhere between eight to ten hours provides the user with around 24 miles use and costs "less than a newspaper in electricity", making it extremely efficient and eco-friendly. Compared to travelling by car, the Segway HT creates fourteen times less greenhouse gases; in fact, it produces no gas-emissions whatsoever, and therefore allows it to be practically used anywhere, including indoors. Unfortunately for Kamen's noble intentions, in 2006 the British government invoked a law that is over 175 years old: the Highway Act of 1835 means that the Segway HT is prohibited to travel on pavements or on roads, effectively meaning that its use is limited only to privately owned land.

Top Exploded view of the Segway HT showing the driving and self-balancing technology integrated into its platform.

Bottom Police patrolling using Segways.

1769/1885
Nicholas Joseph Cugnot/Karl Benz

MOTORCAR

The first successful steam-powered automobile—a tricycle—was built by Nicholas Joseph Cugnot in 1769, and was used for pulling cannons around an arsenal. Despite this, it is usually the Germans Karl Benz or Gottlieb Daimler who are acclaimed with the invention of the modern car. They both designed and made practical models of petroleum-powered automobiles that are recognisable predecessors to the modern day car. Karl Benz was an iron foundry owner, and in 1885 pipped Daimler to the post by creating the first petrol-powered automobile. He began selling his tricycle as the 'patent motor wagon' in 1887. In the autumn of 1888, Berta Benz tested the capabilities of her husband's motorcar by taking it on a 150 mile round trip from Mannheim to Pforzheim. This was the first ever long-distance automobile journey and was not without its problems; Frau Benz had to use a hair clip to unblock a fuel line, repair a short circuit with stocking elastic, and employ a cobbler to fix a broken brake.

In 1893, Benz had developed his ideal model into the first mass-produced car—the Velo—manufacturing and selling 150 in 1895. Unfortunately, Benz stuck to his old designs and never took increasing speed into account, and consequently Daimler and his more innovative designs overtook the flagging Benz until the companies merged in 1926.

1894 saw Emile Levassor and Rene Panhard devise the "Panhard-Levassor", the first motorcar that shared a recognisable aesthetic and mechanical design to the modern day car. They negotiated the rights to use Daimler's high-performance engines, but it was where they placed it that made all the difference. The two-cylinder engine, as well as the radiator, was positioned at the front of the car and enclosed; included were a rudimentary gearbox and friction clutch, and all the power was transmitted to the rear of the car, making it easier to steer.

It was the notorious Henry Ford that introduced the production-line manufacturing of cars. Though first introduced in 1908, his groundbreaking Ford Model-T hit its peak in 1914—with sales passing half a million dollars—a year after Ford introduced assembly belts to his factories. The cars were famously produced at such a speed that they could only be bought in black, since the only paint that would dry quick enough was black Japan enamel paint. Over a 20 year period, 15 million Model-Ts were built and the price fell by 75 per cent. By the end of the twentieth century there were some four million people working in the car manufacturing industry making around 36 million cars a year, and around 500 million cars were being used globally.

From top to bottom Early petrol-powered car, 1880s; Ferdinand Porsche at the wheel of a hybrid-electric Porsche car, 1902; a cycle-car, built by the German engineer Neumann circa 1930.

Opposite Cars in a traffic jam.

1899
Ferdinand Porsche

THE HYBRID

 The hybrid car has a simple ecological objective —to use less fuel and decrease air pollution. Although the first petrol-electric hybrid vehicle was developed over a century ago, no car manufacturer would invest in the concept or mass-produce the cars as they were seen as not commercially viable in a world that did not see the need for them. It took 100 years and a change in environmental awareness for the hybrid to finally arrive, in the form of the Toyota Prius, as a mainstream alternative to the conventional car.

The hybrid concept combines gas power and electric power into one system, increasing the mileage and reducing harmful pollutants emitted by a gas-powered car while simultaneously overcoming the shortcomings of an electric car; the combining of two such energy sources meant that the gas-powered engine could be much smaller than usual, directly impacting its efficiency.

The fist hybrid vehicle was built in 1899 by Ferdinand Porsche. It had a singular gas engine that was used to charge up the batteries of the in-wheel electric motors. Despite being a major breakthrough for the time, there was no need for the Porsche hybrid—known as the Lohner-Porsche—due to low fuel costs, and for this reason the cars were never put into production. The problem of perceived lack of demand has dogged the invention ever since.

In 1974, a whole thirty years before the launch of the Toyota Prius, Victor Wouk produced the first prototype of a modern-day hybrid car made from the chassis of a Buick Skylark; this prototype, gas-electric hybrid used fuel at half the rate of most gas-powered cars and therefore reduced emissions significantly. Although the hybrid car met all guidelines of the US Environmental Protection Agency, it was rejected, mainly by head of Mobile Source Air Pollution Control Programme Eric Tork, who was not a fan of the hybrid concept for the traditional reason of a lack of commercial visibility. He even declared that he would not test Wouk's hybrid, and despite being eventually pressed into doing so, there was a strong belief, subsequently disproved, that whatever the result, the hybrid would not be accepted under any circumstances.

Wouk continued to champion the benefits of the hybrid car, believing that it was the only practical, effective, and—most importantly—readily available technology for the future. He wrote many essays and conducted numerous lectures on the subject. Then, in the 1990s he saw his dreams come true with the release of the Toyota Prius. Revealed in Japan in 1997, and released worldwide in 2001, the model has been greeted with huge industry acclaim and commercial success, setting the course for viable future developments in hybrid and eco-centric transport technology.

Right The Hybrid Toyota Prius.

6000 BC

ROADS AND TARMAC

The first roads originated as historical trade routes, with early travellers and tradesmen impressing pathways into the earth as between disparate historical centers of commerce. Such reasoning explains how the notion of road creation is so antiquated; indeed, it is thought that tracks around Jericho could have first been walked as early as 6000 BC. A more specific early pathway is the Ridgeway in Southern England; thought to be over 5000 years old, the route passes Stonehenge and other sites of historical importance.

Civilisation and cultures advanced; populations increased and so did trade, with more and more people travelling further and further afield to trade in livestock, food, clothing and jewellery; eventually trade became international and a network of semi-surfaced roads were produced throughout central Europe, known as the Amber Routes.

In Crete, a limestone-paved road 30 miles in length was built by the Minoans, running from the city of Gortyna to Knossos. It is the earliest known surviving carriageway and dates back to 2000 BC. It had gutters and was raised in the middle so that rainwater could run off of its surface. Many of the first roads were built for ceremonial purposes; the Chaldeans of Babylon built brick roads that linked royal palaces to temples. But it was the Romans who would be most famously remembered for their roads, being the first to grasp the potential beyond ceremonial use.

The Romans realised that roads were a powerful military tool allowing good lines of communication and trade, and they used concrete to lay a 53,000 mile network, which included around 30 main routes for military purposes, facilitating swift manoeuvring. They were the first master road builders and some of the foundations for their roads were over a metre thick, which is in part why some are still in use today.

Modern road-building began in eighteenth century Britain, purportedly due to the constant state of disrepair afforded by frequent wet weather. Scotsman John McAdam encountered these problematic surfaces when trying to travel around his estate and began developing techniques that would improve roads, leading to the so-called "Macadamised" thoroughfares of the period. The basic principle was to lay a strong foundation of stone block and then layer upon it ever decreasing sizes of stone until the top layer was just gravel; this surface was then pressed down into place using heavy rollers, and water was used to increase bonding. This process was facilitated by the invention of the steamroller in 1865, 30 years after McAdam had died. McAdam and others had built thousands of miles of highway by the time he died, but these roads had one major problem in that they tended to become soft or dusty depending on the weather. A solution was found in sealing the surface using a liquid spray derived from coal tar, which bound the surface and made it waterproof. These roads became known as 'tar-macadam' or 'tarmac' roads.

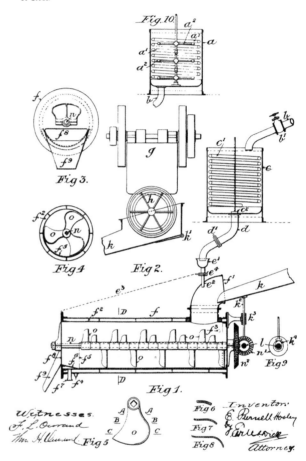

1920
William Potts

TRAFFIC LIGHTS

Left A patent illustration showing a three-way traffic signal, dated 27 February 1922.

The modern day traffic light system of green, amber and red lights was devised by Detroit policeman William Potts in 1920, and was an adaptation of signalling technology that was used on railways at the time. The need for a system of controlling traffic was imperative in the United States during the first decade of the twentieth century as the volume of traffic soared with the advent of inexpensive models of car such as Henry Ford's Model-T.

Railway control systems had been the inspiration and basis for 50 years of road-traffic signalling technology before the arrival of Potts' system. The streets of Victorian London were used primarily by horse-drawn vehicles, requiring an abundance of police to keep traffic moving freely. In 1868, JP Knight, a railway engineer, devised a system that was to be set up outside the houses of parliament. It employed a gas-lamp and semaphore arms operated by a single policeman; when a lever was pulled the arms would raise and rotate a red glass disc in front of the gas-lamp as a warning, while a green glass disc with the arms down would signify a safe passage. However, the system proved dangerous—not so much for the drivers but for the policemen operating it, one of whom was operating the device when it exploded in 1869, causing severe injuries. The system was soon abandoned.

The first electric equivalent, consisting of only green and red lights, was put into operation by the head of traffic in Salt Lake City, Lester Wire, in 1912. A police officer controlled the lights from a shelter nearby, avoiding exposure to the danger of passing traffic and allowing him protection from the elements.

With the progress of technology, traffic-light systems were able to be programmed to change at timed-intervals, and more recently, traffic-systems have become responsive to the flow of traffic, adjusting according to information fed from electronic circuits placed in the road.

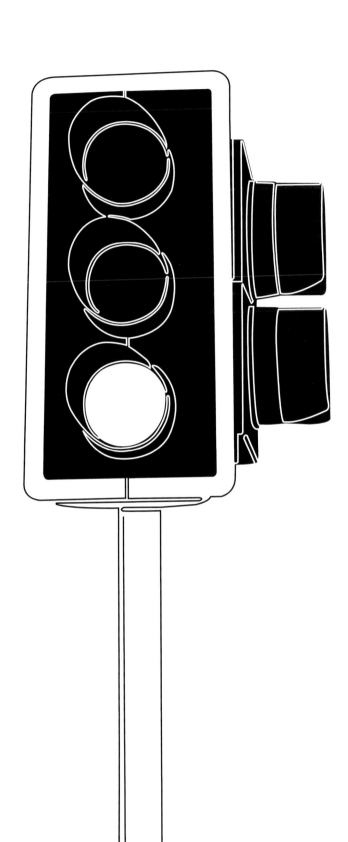

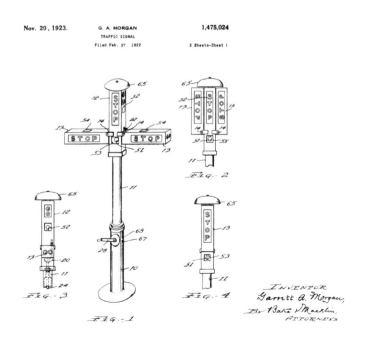

1934
Percy Shaw

CATSEYES

Although electric lamplights were in operation by the early 1930s, most roads, even at this time in the twentieth century, were generally unlit. With the moon remaining the primary light source at night, keeping to the road could prove difficult. Percy Shaw, a road contractor from Yorkshire, saw an interesting solution to this problem via an opportunistic encounter with a cat.

In 1933 Shaw was driving in the dark to his home in Halifax, when his dim six-volt headlamps caught the wide-eyed gaze of a cat perched on top of a fence up ahead; the luminous reflection provided the illumination for solving the hazards of driving at night. Using the cat's eyes as his inspiration, Shaw set about trying to imitate their phosphorescence. The 23-year old prototype used a glass lens and mirror. The finished model, patented in 1934 and put into production a year later, was ingenious not just for its effectiveness as a retroreflective safety device, but also because the Catseye was relatively easy to maintain due to its self-cleaning system. Each individual Catseye had its own moulded-housing that was recessed into the road. When it rained, the housings filled with water, which was then squeezed out and over the lens as cars drove over it and depressed it into the housing. Rubber located at the top of the housing then wiped the lens clean as it rose back out of the ground. This function is still prevalent in the existing design used today.

Shaw's device was largely overlooked; its full potential was not realised until during the Second World War when it became necessary for automobiles to be driven without the use of headlights during periods of blackout so that they would not be seen and targeted by enemy forces. Catseyes in fact proved invaluable during the war and in 1947 James Callaghan, the Labour Party's roads minister, set out a plan that would see them installed throughout Britain on all its main roads. Shaw died in 1976, but his device has continued to save innumerate night-time drivers from injury or death on roads since.

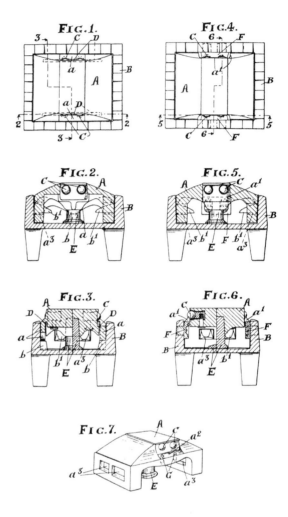

HOVERCRAFT

Below Examples of drawings from Christopher Cockerell's original design specifications, circa 1960.

The purpose of the hovercraft concept, also known as the Air-Cushioned Vehicle (ACV), was to design a boat that did not drive through the water but glided on top of its surface. In reality, the 'hovering' action allows the craft to move over any relatively flat surface. The development of technology that would allow this to happen was first demonstrated in 1906 by a French Count, Charles de Lambert, who attached an aeroplane engine to a flat bottomed boat. The machine effectively hydroplaned over the water's surface at an impressive 38 mph. Italian airship designer Enrico Forlanini engineered a similar machine called a hydrofoil which employed the same physics that produced lift in an aircraft to that of a boat. The hydrofoil had wings that were located under the water and lifted the boat's hull above the water, avoiding drag. Hydrofoils were used around the world as ferries for travelling passengers, and further development saw the 'jetfoil' take to Japanese waters in 1974, powered by a jet engine and able to accommodate 400 passengers at a time.

The invention of the modern hovercraft, however, is accredited to British engineer Christopher Cockerell. His father, in a moment of cynicism, once described his son as "no better than a garage hand", a statement not only discredited by Christopher's invention of the hovercraft but also by his work at the Marconi Telegraph Company from 1935–1951. During these years Cockerell worked with a crack elite team developing radar technology, which Winston Churchill believed to have had a great effect on the allies victory in the Second World War.

He moved to Norfolk soon after and began work trying to reduce the friction produced by the boat's hull coming into contact with the water's surface. The success of Cockerell's design came with his ingenious adaptation to the cushion of air ubiquitous in the machine's appearance. Instead of having just one cushion filled with air, Cockerell believed that using a low-pressured cushion inside a sleeve of high pressure would improve the hovercraft design. In 1955 he tested his ideas by rigging up an empty tin can of cat food inside a coffee tin, airflow was provided by a vacuum cleaner working in reverse, and some kitchen scales replaced the water surface; incredibly, it worked.

A radio-controlled model was used for a demonstration for the British Government's Ministry of Supply. A full machine was created in 1956. The MoS made the development of the technology a secret, but failed to envision a realistic practical use for it. Eventually in 1959 the project was declassified, which allowed for the production of hovercraft for commercial purposes. On July 25 1959, Cockerell's SR-N1 Hovercraft set out on its maiden voyage from Dover to Calais over the English Channel, taking two hours and three minutes. Regular trips across the channel by hovercraft did not begin until 1978, and used a larger model, the SR-N4, which could carry 400 passengers and 55 cars. The main problem with using such large hovercraft was the high running costs because of their voracious fuel consumption and the high level of maintenance necessary to keep them running.

In the present day small hovercrafts are most commonly used for overcoming transitional terrain by the military and for rescue missions because of their incredible versatility. For this reason hovercrafts are not used for ferrying a great deal any more but the technology has been developed for use in other spheres, such as hover-mowers like the Flymo, as well as in some instances of heavy-load transportation, and in medicine where victims of severe burns are rested in a 'hoverbed'.

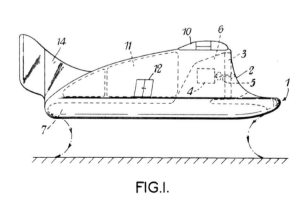

FIG.I.

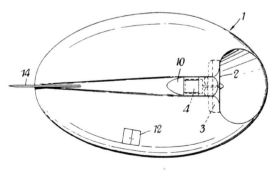

FIG.2.

TRAINS AND RAILWAYS

The nineteenth century and the industrial revolution saw the arrival of the steam locomotive and railway, but it is believed that rudimentary wooden rails were commonly used throughout central Europe by the early sixteenth century. They facilitated the moving of heavy loads from workface to pithead to great effect, whether a wagon was pulled by hand or pony, perfect for using in mine tunnels due to their size and strength. In time, the wooden rails came to be covered with iron strips, which diminished wear and tear, and were eventually made solely from cast iron.

These railway or tram lines were used by horse-and-cart alone until 1804, when a Cornish engineer by the name of Richard Trevithick demonstrated the effectiveness of his steam-powered engine, pulling trucks with a cumulative weight of 20 tonnes along a tram line in Pen-Y-Darren, South Wales. The "iron horse: as it was nicknamed, was an engineering triumph but was nonetheless shunned, as people preferred the traditional horse-and-cart system. This was not only the result of an attachment to pastoral ways; it was also largely down to the fact that under the extreme weight of the engine, the cast iron rails tended to crack, making it an imperfect system.

The rich history of nonindustrial rail transport also developed from mining practises. George Stephenson came from a poor mining background and in 1812 became a colliery engine builder. By 1814 he had designed "The Bulcher", the first efficient locomotive, which was used to haul coal at four mph at the Killingworth colliery in Northumberland. Stephenson set up his own factory to make railway and locomotives, making incremental improvements in their design. His skill was demonstrated when he was asked to build the first public railway line between Stockton and Darlington; the tracks were made out of wrought iron to avoid cracking and it ran at an incredible pace of 16 mph.

Top right The streamlined 20th Century Limited Locomotive, 1938.

Bottom right A steam-powered train in Berlin Ostbahnhof.

Left Cut-away diagram of an early steam engine.

Opposite Workers during the construction of the Siberian railway circa 1895.

Development continued when Stephenson and his son Robert invented the "Rocket" steam engine in 1829, increasing speeds to 30 mph and improving efficiency and power. The Rocket's inaugural journey ran on a line engineered by Stephenson, from the port of Liverpool to the city of industry that was Manchester in 1830. This was the catalyst for over 6,000 miles of railroad to be laid in Britain by 1950, cutting a journey from London to Edinburgh, that would have taken 12 days by horse and cart, to just 50 hours on a steam locomotive. America followed suit soon after the engineering of the Liverpool–Manchester line, creating what was deemed a "wonder of the world" at the time: a 125-mile line. During the 1930s, subsequent railroad lines were built in Russia and in Europe, and Stephenson himself built lines in Belgium and Spain.

With speeds increasing, a train could travel hundreds of metres before stopping, as the brake had to be applied by hand. A young 20-year old rotary steam engine inventor from New York was on one such train in 1866. George Westinghouse set about devising a braking system that meant all brakes on all carriages could be operated at the same time, from the same location, by the use of compressed air pumped through flexible pipes. His system became mandatory law for all steam locomotives in 1893.

A demand for faster travel, and the availability of other sources of energy such as electricity, has allowed the development of locomotive technology and consequently an increase in speeds. Japan's Shinkansen, also known as the "bullet train", was put into operation in 1964, and can travel at a top speed of 310 mph. Magnetic levitation, or mag-lev, locomotion, whereby trains actually float above the track and are guided by electromagnetic waves, was developed to further potential in this area. Trains encompassing the technology can reach top speeds of 370 mph without a motor, making the journey extremely efficient and economic. They have been in limited use in Japan since the 1970s.

Top to bottom The Trans-Europe Express; a US high speed train; a Shinkansen train in Nagano, Japan.

circa 1490
Leonardo da Vinci

ORNITHOPTER

In the ancient Greek myth, Daedalus and his son Icarus, having been imprisoned in a tower by King Minos of Crete, speculated on how to escape. Reasoning that the King could rule the land and the sea, but not the heavens, Daedalus proceeded to create a set of wings for both of them out of bird feathers bound together by wax. By emulating the flight of birds, the pair's plan was a great success, until the unruly Icarus flew too near to the sun, melting the wax that held his wings together and sending him plummeting into the sea to drown. Although the tale ends in tragedy the principle of imitating the motion of birds in flight to create flying machines is an idea that has remained and has been developed for centuries; such somewhat fantastical machines are commonly referred to as Ornithopters.

Leonardo Da Vinci, in around 1490 after studying the flight of birds, had much the same idea as Daedalus but quickly realised that a pair of wings attached to a human's arms would not be sufficient for aviation, humans being too heavy and lacking the necessary strength. He did, however, produce diagrams and sketches of a machine that would follow this principle but more closely resembled a modern day helicopter. Da Vinci was right in presuming that it was practically impossible for a human to fly using the same mechanics as a bird; many ornithopters have been built in recent times based upon the wings of birds and insects, but invariably they have been the same size as their flesh-and-blood counterparts. Practical applications of ornithopers are actually made to look like birds and insects as well. An artificial hawk under the control of an operator has been used in Colorado USA to keep the endangered Sage Grouse grounded in order to capture it for research. Prospectively they could be used by the military for spying, and in fact Paul B MacCready of AeroVironment Inc. has invented a remote-controlled ornithopter the size of a large insect that has the potential to carry out spy missions. Ornithopters are more commonly used by hobbyists and flight enthusiasts in the form of lightweight models powered by rubber bands.

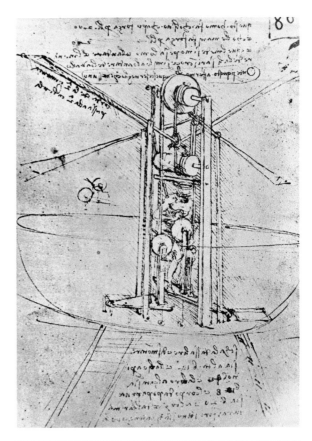

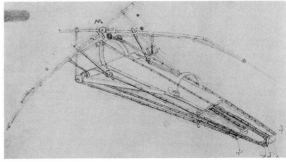

Top An illustration by Leonardo da Vinci of a rudimentary flying machine, entitled *Drawing of a flying machine with a man operating it*, 1488.

Bottom Ornithopter designed by Leonardo da Vinci, circa 1486.

1782
Joseph and Jacques-Etienne Montgolfier

HOT AIR BALLOON

Invented by the French brothers Joseph and Jacques-Etienne Montgolfier, sons of a wealthy paper merchant, the brothers had access to the money and technology needed to make manned flight a reality.

Joseph was familiar with new discoveries made within the field of the chemistry of gases, and more specifically the isolation of hydrogen, a method devised by Henry Cavendish. Hydrogen is lighter than air—also known as having a 'negative weight'—and if harnessed could be used to lift things from the earth's surface; however, it was extremely hard to isolate in the quantities that Joseph would need it in order to power an air balloon. Sitting in front of his fire in the November of 1782 Joseph noticed embers rising on a wave of hot smoke. He realised that the readily available and simple to prepare hot air could act buoyantly in a similar fashion to hydrogen. Building a lightweight framework, covering it with taffeta fabric and leaving a small hole on the underside, Joseph had built the first miniature hot air balloon. He immediately wrote to his brother, forecasting: "Get in a supply of taffeta and of cordage, quickly, and you will see one of the most astonishing sights in the world."

Joseph and Etienne endeavoured to construct a version nine feet in diameter in mid-December. Even with hefty restraining cords and only half-filled with hot air, the balloon soared into the sky and came to land over a mile away. In 1783, after some modifications including using sack cloth with three thin layers of paper to replace the heat-leaking taffeta, and increasing the balloon's size to a massive 35 feet across (which allowed it to hold one ton of air), the Montgolfier brothers gave a public demonstration of their invention in the marketplace of Annonay. On a rainy day, heated by a brazier, the balloon reached a height of 3,000 feet. After travelling a mile-and-a-half it crash-landed in a vineyard and caught on fire. After the rather hazardous landing in July, the brothers spared human passengers the very real danger of being burnt alive, and instead sent up a rooster, a sheep and a duck into the clouds for eight minutes in September, with Louis XVI and Marie Antoinette watching. The Montgolfiers became famous overnight. However, it is the maiden voyage of man into the sky above that will forever be synonymous with the same name. On November 21 of the same year, passengers Pilâtre de Rozier and Marquis Arlandes sailed over Paris in a Montgolfier balloon at an altitude of half a mile and landed 25 minutes later completely unharmed.

On December 1 of the same year, fellow Frenchmen Jacques Charles and Nicholas Robert took to the sky in a hydrogen-powered balloon, combining Charles' knowledge of making hydrogen and Robert's new method of coating silk with rubber to create the balloon. The voyage was predated by a rather more disaster-laden affair in August. The hydrogen was produced by adding sulphuric acid through a tube and bunghole into a barrel of iron filings. It took 500 pounds of sulphuric acid to produce enough hydrogen to fill the balloon to capacity. Problems arose when the heat produced by the chemical reaction evaporated some of the sulphuric acid, causing it to condense on the balloon and burn through the lining. These faults were ironed out for its unmanned maiden voyage, but other problems were encountered: the balloon shot upwards and was out of sight in two minutes, continuing to rise until it burst due to low air pressure at high altitudes. It landed fifteen miles away on agricultural land. Still full of hydrogen and bouncing along the ground, curious peasants thinking it was alive stabbed it with pitchforks and then tied it to the tail of a horse to be carted off.

1848
John Stringfellow

AEROPLANE

It is generally agreed that American brothers Wilbur and Orville Wright were the first to fly a 'heavier-than-air', powered aircraft for a sustained period of time. There had been 100 years of development and a plethora of inventors and engineers desperate to become the first to take to the sky before the Wright brothers' maiden flight on 17 December 1903, but the coupling of extraordinary engineering skills and precise methodical research over half a decade made their Wright Flyer No 1 a success. A wholly original propeller system had to be designed and an engine had to be built to power it. However, although the Wright brothers are worthy of the regard bestowed upon them in the world of aviation, their success was, in part, built upon the incremental developments of others, and the failures of less fortunate engineers.

The first milestone of this lineage is located at the door of the veritable godfather of the aeroplane, British inventor Sir George Cayley. Cayley devoted much of his life to the research of aeronautics, producing theories surrounding the principles of aerodynamics and the configuration of a modern fixed-wing aeroplane. His successes culminated in 1853 when he persuaded his coachman to be carried into the air for a short distance by a full-size glider that he had built. This was not however powered flight. The first patent for powered flight was granted to British inventor William S Henson's "Locomotive Apparatus for Air, Land, and Water" in 1843. Henson worked alongside good friend and maker of miniature steam engines John Stringfellow to build a steam-powered aeroplane with a six-metre wingspan, but it was too heavy to fly and Henson soon lost all interest in the endeavour. Stringfellow persevered after Henson's exit and built a lightweight version of their original plane with only a three-metre wingspan, which flew a modest distance of ten metres in 1848. It was progress nonetheless.

Another set of brothers from Germany, the Lilienthals, made very important inroads in the world of manned gliders, learning how to manipulate the flight of the machine through the shifting of body weight and changing the shape of the wings via the use of cords. Between 1891 and 1896 the brothers recorded every single one of their 5,000 flights meticulously which served as an invaluable resource for pioneers to come. During 1896, American Professor Samuel P Langley built another steam-powered model aeroplane that managed to stay in the air for 90 seconds while flying a distance of more than half a mile. With a development grant from the US War Department of $50,000 the professor built the Langley Aerodrome, a petrol-powered machine that on its first test flight on 7 October 1903 snagged on take-off and promptly plunged into surrounding waters.

Otto Lilienthal tragically died in 1896 when he lost control of his glider and crashed. Reading his obituary whilst nursing Orville back to health from a bout of typhoid, Wilbur Wright inadvertently rekindled the pair's childhood fascination with aviation. Once Orville had recovered, the brothers undertook in-depth studies of the works of Cayley and Lilienthal, and built and tested a full-size glider in 1900. After further rigorous experiments and research regarding wing shape and size, Wilbur and Orville tested their Flyer No 1 on 17 December 1903. Its wings were a wooden-ribbed construction

Top The first successful flight by the Wright brothers, 1903.

Bottom The Wright brothers experiment with manned gliders, circa 1900.

covered in cotton cloth with a span of 12.2 metres. The pilot lay in the middle of the biplane with a 12 horsepower petrol engine mounted beside him, controlling the machine via a cradle wrapped around his waist. A cable attached to the wings' tips responded to the turn, pitch and roll of his body movements. Orville won the toss of the coin and thus the opportunity to put his life on the line in the name of aeronautic progress; and at 10:35am the Wright Flyer No 1 left the ground with Orville inside, and proceeded to fly at a height of three metres for 12 seconds covering a distance of 37 metres, all of which Wilbur also covered running alongside the airborne machine and his brother. After three more successful flights, the most successful being a 59 second flight covering 260 metres with Wilbur at the controls and with a strong headwind, the Flyer No 1 got caught in one such gust of wind as overturned and destroyed. No one was hurt and the brothers awoke the next day to find their achievements documented in the local newspaper.

Questions surrounded the authenticity of this particular powered take-off, essentially because it was a manned glider powered by strong headwinds with the engine reserved for steering purposes. But in the summer of 1905, the brothers justified their leading positions in the field by making over 40 flights in their Wright Flyer No. III. Without the help of headwinds the Wrights were able to stay in the air for up to 40 minutes at a time, producing turns, figure of eights, and circling at speeds of 35mph. Concerned that their rivals would steal their design, the Wright brothers remained grounded until patents had been filed, and until they had sold the licensing rights to both the US army and a French company in 1908.

Improvements continued but every successful journey was counted as literally death-defying, not least Louis Bleriot's crossing of the channel in an improved monoplane in 1909 in 37 minutes. Perturbed by France's new military potential, one commentator remarked that "England is no longer an island", whilst politician Davis Lloyd George warned that "flying machines are no longer toys". And he was correct: with the advent of the First World War, aircraft development increased rapidly; during this time machine guns were mounted to shoot through the propellers during dog fights and planes were made to be stronger, faster, and more manoeuvrable. 1914 saw Hugo Junkers devised the first all-metal monoplane with low, cantilever wings, that anticipated future designs and was adjusted for military use in 1917. The first cross-Atlantic flight took place in 1919 when British aviators John Alcock and Arthur Brown flew a stripped-down Vickers Vimy bomber for 16 hours from Newfoundland to Ireland— "sometimes upside down in dense, icy fog" according to the New York Times—and in 1921, the first pressurised cabin was installed in a British aircraft, the De Havilland DH4, so that the passengers could fly at high altitude. But by the mid-1930s the piston-driven aircraft had reached its limits and made way for the much more efficient jet engine.

Opposite The Airbus A350.

Top to bottom The first airplane to cross the Atlantic, 1927; *The How and Why Wonder Book of Airplanes and the Story of Flight*; a view from the promenade of Nagoya Airport, Japan.

JET ENGINE

The idea of using the jet engine for powering planes was only truly developed with the advent of the Second World War. In 1930 Frank Whittle, a Royal Air Force cadet, had registered the first patent for a jet engine, but Whittle's invention was greeted with little encouragement and even a smaller amount of funding. It was German aviation engineer Hans Ohain who, despite receiving his patent four years later than Whittle and in complete ignorance of the British engineer's progress, received the financial backing of plane maker Ernst Heinkel, in 1936, that allowed the Germans to become the first nation to fly a jet engine. The Heinkel He 178 flew in August 1939, and Heinkel proclaimed that the jet engine's "hideous wail" music to his ears. Whittle eventually saw his Gloster Meteor take flight two years later and then saw its entry into allied service in 1944, a tool for destroying German V1 rocket bombs.

The Second World War kickstarted the development of jet engine technology in the UK; the British built the first turboprop, the Vickers Viscount, in 1948, based upon the 1945 designs of inventor AA Griffith, who had the idea of using jet exhaust to power plane propellers many years earlier, in the 1920s. In 1949 the De Havilland Comet I, the first jet airliner, took its maiden voyage using its four engines to fly at a speed of 490 mph; this was the forefather of the ubiquitous Boeing 747, the first 'jumbo jet', which carried 500 passengers close to the speed of sound in 1970.

Concorde, like the Boeing 747, was not only a touchstone in aviation technology but also one of the most recognisable design icons of the modern world. Furthermore, it saw a collaboration between British and French teams—the former made the engines and electronics, whilst the latter constructed the afterburners and hydraulics. The aircraft could carry 100 passengers at speeds over twice that of sound, cutting travelling time across the transatlantic to less than three hours. But Concorde never became the commercial success that it was intended to be; only 14 models were made from its inception in 1976 because the advantages of flying at such high speeds came at a cost. It consumed huge amounts of fuel and was only permitted to break the sound barrier over the Atlantic due to the amount of noise created. Because of this, subsequent designs centred around improving the efficiency of planes rather than simply the speed, and designers began looking back to previous ideas to achieve this.

The development of new ultra-light and strong composite materials in turn allowed further developments in plane technology; American Burt Rutan's Voyager of 1986, for example, was able to fly nonstop around the world without refuelling. The BD-10, the first of the private supersonic jets, was invented by American Jim Bede in the early 1990s. A twin-engine two-seater plane that weighed just 250 pounds and measured less than nine metres in length, it could reach speeds of over twice that of sound and climb 6,000 metres per minute. Its wing design and light yet strong materials made this possible.

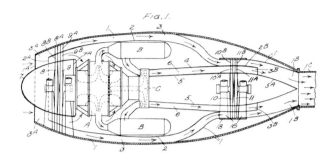

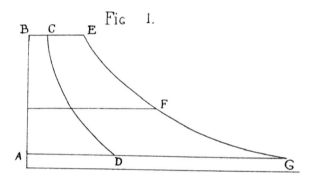

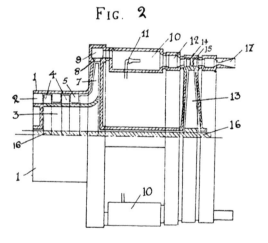

Above Illustrations variously extracted from Whittle's design specifications, relating to the first design for gas turbine jet engine propulsion.

ROCKET PROPULSION

The development of rocket propulsion technology began in the Soviet Union, through the tests and calculations of schoolmaster Konstantin Tsiolkovsky. Tsiolkovsky constructed wind tunnels in order to test the effects of air friction and spent years of his life calculating what would be needed to create a velocity great enough to escape the earth's gravity. He eventually published his theories in 1903, the same year that the Wright Brothers took flight.

These theories sparked the imaginations of numerous great minds around the world during the early part of the twentieth century. One such mind was that of American physicist Dr Robert Hutchings Goddard, who had been obsessed with reaching the moon from a young age after taking inspiration from the tales of Jules Verne and HG Wells. After years of experimentation using static solid fuels, Goddard concluded that they were too hard to control and lacked efficiency; he deduced that the use of liquid fuels was necessary, theorising that if hydrogen could be siphoned into a combustion chamber sufficiently quickly and burnt with liquid oxygen, enough power would be produced to hypothetically reach the moon. In 1919 Goddard published a report on his findings called *A Method of Reaching Extreme Altitude*; the *New York Times* lampooned his writings, stating that travel to the moon was impossible because there was no atmosphere on which to push against. They claimed that Goddard lacked "the knowledge ladled out daily in high schools".

Enraged by the comments, Goddard set about making his rocket in secret. In 1926 he built a rocket powered by petroleum spirit and oxygen; it reached a height of 60 metres, not quite enough to break the earth's atmosphere but the principle the rocket was based upon was sound nonetheless: the oxygen carried on board would enable fuel to burn in the airless void of space, thus creating propulsion through its exhausts and allowing a craft to move through space. Nine years later Goddard sent one of his rockets 2,286 metres into the air and had by then long been attaching instruments to his rockets including barometers, thermometers, and cameras: he called his missiles "Nells" but the US government and Army largely ignored Goddard's progress.

Top Take-off of the space shuttle Atlantis.

Bottom An example of a rocket reaction engine.

Opposite A replica V-2 rocket, at the Peenemünde Museum, northeast Germany.

Meanwhile in Germany, 22-year old Wernher Von Braun, a bright young assistant to visionary professors Herman Oberth and Walter Hohmann, designed his prototype 'A-2' series with great success in 1934; they were long-range rockets fuelled by ethyl alcohol and liquid oxygen. His greatest moment came with the invention of the A-4, whose first wholly successful launch occurred on 3 October 1942. The 12-tonne rocket reached a height of 50 miles and could travel a distance of 200 miles at five times the speed of sound. This revolutionary invention became the infamous V-2, or the Vengeance Weapon 2 used in the Second World War, and was loaded with explosives to become the first strategic missile. Towards the end of the war, a number of V-2 rockets were captured by the USA as the Germans were defeated, at which point Goddard was eventually able to see the inner-workings of the missile and recognised it as his own handiwork. It is believed that the Nazis had read and appropriated all of his papers and 200 patent applications.

The V-2s were eventually utilised for Goddard's intended purpose, with instruments attached to them that allowed for the scientific exploration of the upper atmosphere, and prospectively into space. However, Goddard died in 1945, never living to see his dream come true. By the 1950s the supply of German V-2 rockets had run out, but the USA had also attained the specialists behind German rocketry, including Braun himself, whose ideas and concepts were used as the basis of the Americans' ventures into space.

However it was the Soviet Union who launched "Sputnik", a 61 cm, spherical artificial satellite, into orbit via rocket propulsion on 4 October 1957. The Korolev rocket that delivered Sputnik into space used 514 tonnes of thrust at lift-off and was the most powerful rocket of its time.

1848
James Bogarde

THE SKYSCRAPER

Two developments were needed in order to erect skyscrapers: one was the use of cast iron columns and girders in place of masonry, and the second was the invention of the elevator.

American builder and inventor James Bogarde erected the first such structure in 1848. His cast iron building was a five-floor factory in New York and was a pioneering piece of engineering. Masonry would not work, since to build tall buildings it would be necessary to give the walls on the ground floor an unfeasible thickness, as they would have to bear the load of the whole building. With iron and later steel frames, the weight of the building was diffused throughout the whole structure and a lightweight masonry curtain was attached for protection from the elements.

When the city of Chicago was destroyed in a blazing fire in 1871 the city planners seized the opportunity to rebuild their city bigger, better and taller. The ten-storey Home Insurance Company Building was completed in Chicago in 1885 and it was the German-born architect Dankmar Adler that liberated the skyscraper's potential for height even further by devising the 'caisson' foundation—a cylindrical excavation filled with concrete.

Yet still there were factors that limited the potential height of these constructions. Elisha Otis invented the hydraulic safety lift in 1852, but it was only effective up to about 20 storeys. It was not until 1903, when his company perfected the electric elevator by offsetting the weight of the car with a sliding counterweight, that these buildings began to truly scrape the sky. This first phase of skyscraper construction culminated in the building of New York's 55-storey Woolworth building that remained the tallest building in the world until the Empire State Building was erected in 1931. The Empire State remained the tallest building in the world for 40 years until the completion of the World Trade Centre and was the last of the big steel and concrete structures—as modern skyscrapers were now clad with large sheets of glass to create a huge mirrored curtain. Stronger steel was developed that allowed for just an exoskeleton to be made, freeing up the internal space of the building and allowing the structure to sway in high winds in order that the pressure did not become too much and the building topple; modern day skyscraper design includes the use of shock absorbers in their foundations in order to be able to withstand the movement caused by earthquakes.

The Woolworth Building, New York, under construction, 1912.

**1777
Samuel Miller**

CIRCULAR SAW

The idea for a rotating circular blade first came to Southampton sail maker Samuel Miller in 1777 when he effectively made the first table saw, allegedly powered by wind and manual spinning, whereby logs were pushed into the rotating blade and cut down into strips of timber. Miller used a small circular saw that suited his own needs, but in 1813 protestant Shaker-sister Tabitha Babbitt was the woman to introduce the large circular saws into the sawmills of her community in Massachusetts. Babbitt, who worked as a spinner in the spinning house of the Harvard Shaker community, saw that the men cutting wood for roof shingle with a large double-handed saw at each end, wanted to find a way of easing this labour-intensive process. She attached a tin disc notched with teeth on to the circumference of her spinning wheel to create a circular saw.

The community promptly set about constructing a large spinning saw for the wood to be pushed through the sawmill. These large circular saws were powered by the flow of water of an adjacent river, using a watermill to produce the power needed. Nowadays all circular saws are powered by electricity and come in various forms including handheld, bench saws and chop saws. The principle of a circular blade also applies to grinders and diamond cutters that can cut through metal and concrete.

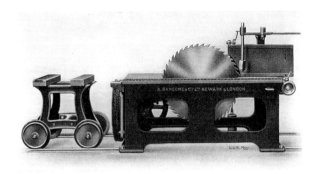

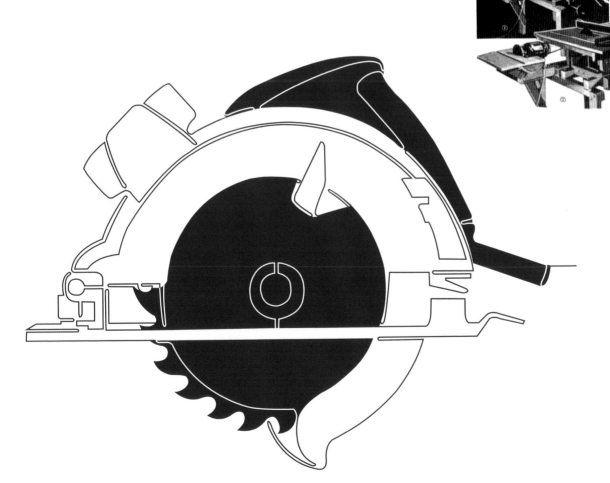

WINDOW PANE

Glass is one of man's earliest inventions. The tale, recorded by Roman historian Pliny (AD 23–79) states that in 5000 BC Phoenician sailors set up camp on the sandy beaches of Syria; not having any rocks to hand to support their cooking pots, blocks of nitrate were used instead. In the morning when they awoke, the sun was gleaming off a piece of glass formed by the fusing of silicon in the sand and soda in the saltpetre. Although there is no hard evidence to validate this story, glass beads of an Egyptian necklace have been found dating from 3500 BC, so it is very possible that these sailors had discovered the formula for producing glass half a century earlier. The processes of molding, shaping, cutting and producing glass established from these early days of antiquity mean that its uses are now varied, from making beautifully crafted and coloured jewellery and ornaments to making huge panes of glass for enormous skyscrapers.

Rudimentary, glass windows have been found in England during excavations dating back to the seventh-century Anglo-Saxon period. Early methods were based around a process of pouring molten glass into a mould, leaving it to partially set and then drawing out a hollow, cylindrical shape, which was then cut down one side, heated so that it was malleable, and rolled and flattened into approximately the right form and thickness. By the early twentieth century a long, wide, flat sheet could be pulled from the molten glass mould. Invariably the sheets of glass had imperfections and ripples that would distort the light shining through the windows during the day, which was not too much of a problem for domestic use but became a major problem when large sheets of glass were needed for shop fronts; it took many hours to grind and sand the imperfections out of the glass, and then polish it to produce the desired 'plate glass'.

It was English firm Alistair Pilkington who already having had 150 years of glass-making experience, devised an ingenious process for producing glass panes in 1959. The process took seven years to develop and relied upon the natural properties of two materials: their so-called 'float glass' process involved floating molten glass over a bath of molten tin to produce a completely flat piece of glass, extremely uniform in thickness. The process made glass production much quicker and cheaper and the quality of the glass was almost flawless. This method was developed further with the advent of the 'electro-float' process in 1967, where the surface of the glass was infused with metallic particles and by using varying levels of heat and light the glass could be altered to any degree desired.

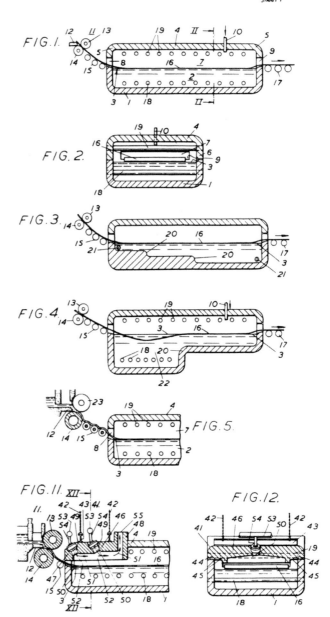

Opposite An early stained glass window.

Right Patent illustration of float glass manufacturer depicting the roller system.

CEMENT

Cement plays a fundamental part in the modernisation and expansion of the modern urban environment. Derived as a binding agent—a material that would set and harden independently, fixing stones or other aggregates together into a solid formed mass—the primary uses of cement are in the production of concrete and mortar.

A ubiquitous element of the every day, concrete is a signature material of the building industry, used for patios, driveways, basements and skyscrapers. The annual global production of concrete is somewhere around 1.25 billion tons, with its universal appeal lying within its base components of some of the most abundant raw materials available: limestone, clay, shale and sand.

Although it is uncertain where exactly cement has its origins, it is exemplified in its earliest forms by the work of Egyptian builders on the Great Pyramids. The Assyrians and Babylonians used similar methods of cement construction, but Roman engineers advanced this, developing hydraulic cement instilled with considerable durability. Two of the most notable architectural punctuations of the Roman landscape—the megalithic dome of the Pantheon and the Baths of Caracalla—were undoubtedly conceived using such a cement and as archaeologists have postulated, this material was a successful blend of slaked lime and pozzolana—a volcanic ash dispersed from Mount Vesuvius.

With a surge in construction at the start of the Industrial Revolution, modern hydraulic cements made their way into existence. Primarily formed from hydraulic-lime, this type of cement was particularly effective for creating renders to finish buildings in wet climates, producing concrete in shipping industries and for the straightforward production of an incredibly robust concrete. The most prolific expansion of the cement industry came from Britain in the 1780s with the development of James Parker's "Roman Cement". Produced by burning minerals rooted in clay deposits containing both clay minerals and calcium carbonate, which were then ground into a fine powder and combined with sand. A mortar was then produced that would set in five to 15 minutes.

With the success of this "Roman Cement" came an increased output of products from rival manufacturers, each vying to produce a stronger, more rapidly strengthening material.

Later, James Frost produced what he identified as "British cement", finally obtaining a patent in 1822. James Aspdin patented a similar material in 1824 which, given its colour resemblance to Portland stone, he christened as "Portland cement".

However, both Frost and Aspdin struggled to contend with lime and pozzolan concretes. The incredibly fast setting time of Portland Cement made coverage and placement over large areas very difficult, and combined with its low early strengths, necessitated its swift re-development. In the early 1840s Aspdin's son William finally produced the Portland Cement that exists today. Adding alite—Tricalcium Silicate—to his mix, Aspdin greatly improved the strength of cement in its early stages of setting. Although his modification of the mix initially placed great financial demands on manufacturing companies, Aspin's innovation gained economic leverage because it was the only product of its sort in manufacture: Portland Cement set reasonably slowly and developed its strength rapidly. As construction rates expanded and the industrial boom intensified, demand for concrete grew with similar speed, and Portland Cement has prevailed as the primary trade leader to date.

SYNTHETICS

The first synthetic, or non-natural material, was invented by Charles Goodyear and patented in 1844. Known as "vulcanite", it was attained by heating natural rubber with sulphur. It could be shaped and hardened, but was always black in colour unless painted. English chemist Alexander Parkes set about improving the material so that the substance could be naturally coloured. He achieved this by 1862 and his newly dubbed "parkesine" synthetic won a medal at the International Exhibition in London in the same year.

Parkesine was derived from cellulose nitrate, which was then dissolved in a solvent such as ether. When the solvent evaporated it produced a residue that was hard but also malleable and ductile, as well as being waterproof. This substance was heated, rolled and pressed into shape and then coloured in whichever shade was required; it was the first thermoplastic. However groundbreaking his invention, Parkes' business had gone into liquidation by 1868 because although it could be produced to a better quality than vulcanite, parkesine was only cheaper if its quality was compromised during production.

It was two American brothers, John and Isaiah Hyatt, who took the next pioneering steps in synthetics. They used camphor —extracted from the Laurel tree—as the plasticiser in cellulose nitrate, and this produced what was termed "celluloid" or "synthetic ivory". It attained this nickname because it was first utilised in the production of billiard balls, the popularity of which at the time led to a massive slaughter of elephants for their tusks. This declined sharply with the production of celluloid. It became the first mass-produced plastic and was used in manufacture of myriad products, including toys, knife handles, collars, cuffs, and photographic film. This continued through to the mid-twentieth century, at which point less flammable and safer plastics with greater durability were being produced.

These pioneering steps in the production of the first synthetic materials, using a natural substance as a starting point, eventually gave way to entirely artificial fibres such as nylon and modern-day polymers; the so-called "plastic age".

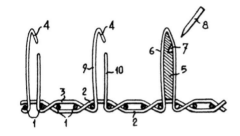

721,338   COMPLETE SPECIFICATION
1 SHEET   *This drawing is a reproduction of the Original on a reduced scale.*

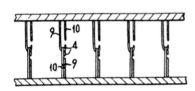

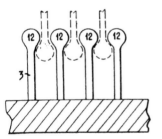

Opposite Nylon fabric.

Top Velcro specification.

Bottom Vintage celluloid pool balls.

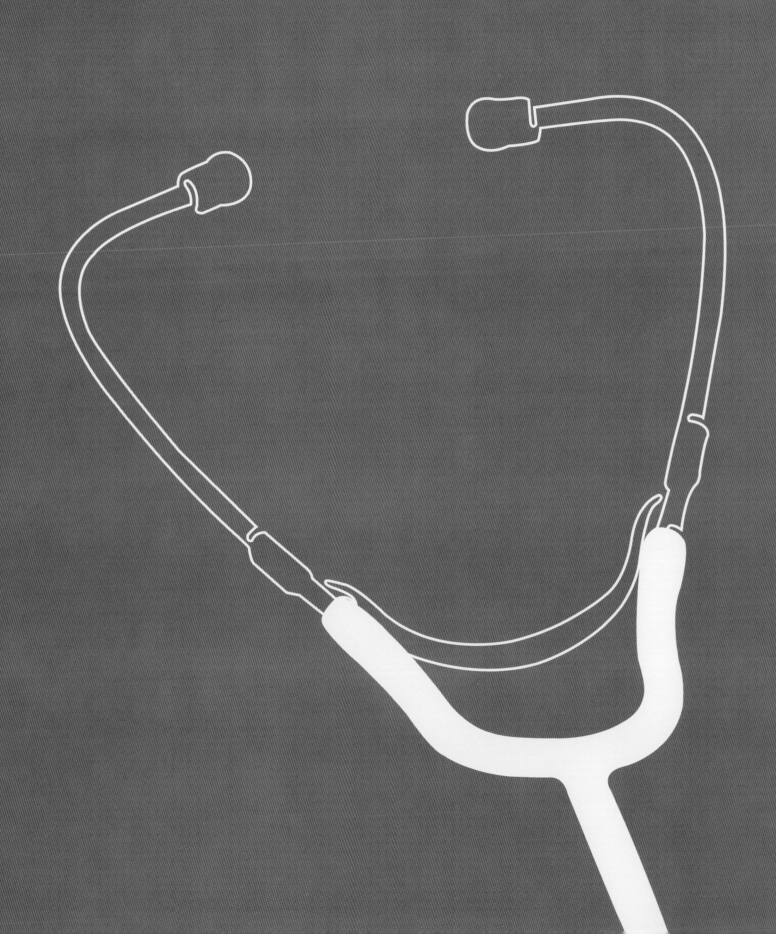

medicine

A child receives inoculation at a 'vaccine farm' around 1944, USA.

VACCINATION

Infectious diseases have been the cause of millions of deaths throughout history. Between one third and one half of the population were killed in Europe alone during the Black Death of the mid-fourteenth century, whilst an influenza outbreak in 1918 killed up to another 50 million people worldwide. However, some people, although being in close contact with the infected, failed to contract these diseases. It was commonly thought that these people had accidentally protected themselves due to being previously exposed to a weak bout of the disease before it had escalated into a pandemic.

By the early eighteenth century doctor's were already advocating injecting people with the very disease that could, potentially, kill them. For example, people would be injected with a mild case of smallpox in the hope to build up an immunity to the disease, but there was a very real chance that this type of inoculation could end in death as most contracted a more virulent strain than intended by physicians. The development of vaccination stemmed from the inspired experiments of Edward Jenner, a surgeon from Gloucestershire. Upon hearing claims that individuals who contracted cowpox, a relatively harmless virus, developed an immunity to the lethal smallpox, Jenner set about the faintly unethical task of testing his theories on a young boy, James Phipps.

In May 1796 Phipps was inoculated with matter extracted from the fingertip of a dairymaid who had been infected with cowpox. Six weeks later, the boy's cowpox-induced fever rescinded, Jenner injected the boy with a potentially lethal dose of smallpox. Fortunately, his predictions were correct and the inoculation had allowed Phipps' immune system to develop a resilience to the more aggressive smallpox virus, almost certainly saving his life.

Jenner thus saved the lives of millions of people in his time and within future generations, as vaccination still remains the primary defence against infectious diseases.

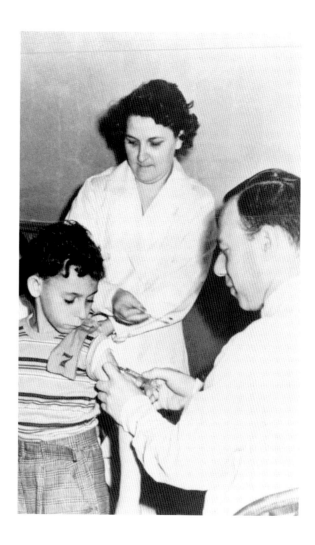

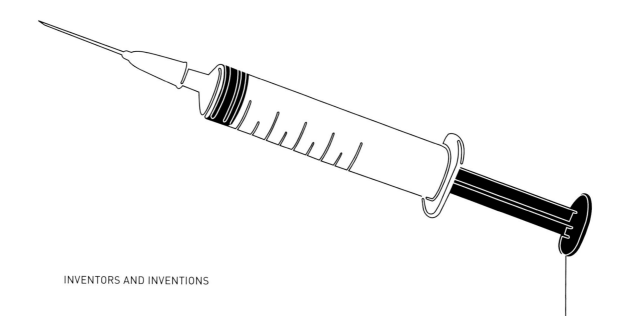

1920s
Earle Dickson

PLASTER (BAND-AID)

Josephine Dickson married Earle Dickson, a cotton buyer for the Johnson & Johnson Company that manufactured gauze and adhesive tape, in 1917.

After accidentally cutting her finger whilst cooking dinner, Earle's makeshift solution was to cut a length of bandage tape into strips, attaching them half to the sides of a small piece of his company's sterile gauze, and half to his wife's skin. The small, improvised bandage stayed in place and did not come off when in use, protecting the small wounds perfectly. From then on, whenever the household saw a domestic accident, ready-made bandages were on hand for her to use, only taking a matter of seconds to apply.

Earle told his colleagues at Johnson & Johnson about his customised bandages, with the result that the company was soon making the ubiquitous repair patch and selling them on a small scale under the name of "Band-Aids". Sales were slow at first but the plaster's popularity soared after an ingenious publicity stunt that saw Johnson & Johnson provide Boy Scout troops with free band aids; four years later, in 1924, the company installed machines for mass producing the new product in various sizes, and began selling pre-sterilized versions in 1939. Vinyl replaced cloth as the Band-Aid's key production material. Earle Dickinson was eventually made vice president of Johnson & Johnson on the back of his invention's popularity.

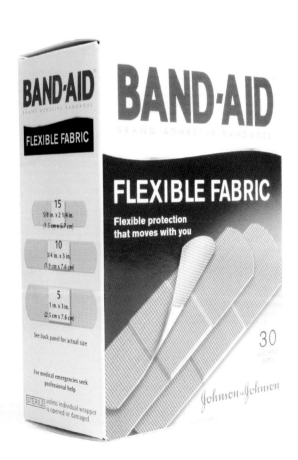

SPECTACLES

Below Theodore Roosevelt wearing a C-bridge type pince-nez glasses, popular in the nineteenth century.

Opposite bottom left Scissors glasses.

Opposite bottom right Snellen chart used to test eyesight.

It is generally accepted that spectacles were first improvised and designed in northern Italy in the last quarter of the thirteenth century, though it is also cited that Marco Polo is said to have witnessed an elderly Chinese woman using a pair at a similar time, whilst travelling the Orient. One of the first references to the use of glass lenses for improving eyesight came from an English friar Roger Bacon, residing in Paris, whose *Opus Majus*, published in 1266, explained the magnifying effect of convex glass that could be utilised as a corrective lens. Though Bacon never developed his theories, he was still met with suspicions of sorcery at the time; ideas pertaining to optical were deemed as unnatural and therefore illegal. This shadow of suspicion shrouded the use of spectacles in the latter part of the thirteenth century, and a means of dispelling this taboo and legitimising their use was required. One of these was elementary: that optical devices were based in science, discovery and geometry; another was slightly more tenuous and had its foundations in Christianity: the degeneration of eyesight was a long-term effect of the Fall of man, and more specifically Adam from his perfect state. By donning spectacles the wearer was only obeying his impulses in recovering Adam to his former state of grace.

Friar Giordano Da Rivalto, celebrated the ingenuity of mankind in a Lent sermon of 1305, declaring "it is not twenty years since there was found the art of making spectacles" dating the use of spectacles to around 1285. The spectacles of this time consisted of convex lenses in order to correct presbyopia (long-sightedness) so that people could read and write with comparative ease past their middle-age. Glass lenses to correct myopia (short-sightedness), improving sight at distances were not produced until a couple of centuries later, at around the same time, 1452 to be exact, the printing press was invented and inexpensive spectacles began to be produced to cash in on the proliferation of reading material and increase in the reading population. The rich upper classes would wear eyeglasses with a frame of gold or silver, whereas the rest of society would wear eyeglasses mounted in wood, bone, leather, or light steel or tin.

It was not until the 1600s when a rigid frame that bridged the nose was used rather than the traditional riveted design. In the early 1700s and 1800s scissor glasses became very popular, hand-held spectacles that hung around the neck in the shape of a pair of scissors, worn by such iconic figures as Goethe, Washington and Napoleon. The late 1700s saw Ben Franklin invent bifocals, and ornate prospect glasses, classy lorgnette and miniature spyglasses were de rigueur in Pre-Revolutionary France; whilst the 1800s saw the lime-light shine upon the iconic Pince-Nez. The latter part of twentieth century brought with it many different styles of spectacles, largely due to the development of the plastics industry.

Though glass contact lenses were invented in 1887 by FE Muller, the twentieth century saw the inception of the plastic lens industry. In 1938, the first 'scleral' lenses made from solid Perspex were made. Covering the whole eye, they were hugely uncomfortable due to their inability to let the eye breath. Soon after, adjustments were made whereby a tiny hole was drilled though the centre of the lens, allowing them to be worn for much longer periods of time without compromising vision. 1948 saw a shrinking of lenses to cover only the cornea, making them easier to fit and cheaper to produce. Despite this, the breathability problems inherent with using Perspex prevailed. In 1960, Czech polymer chemist Otto Wichterle proposed the idea of using a soft, jelly-like, water-absorbing lens in place of the previous plastic. The concept was bought by American optical company Bausch and Lomb who eventually launched the product in 1972. Exactly one hundred years on from Muller's creation of the glass contact lens, the first ever one-day disposable contact lenses were invented.

SHATTERPROOF GLASS
USED IN SPECTACLES

HAMMER blows cannot shatter the glass in safety spectacles recently developed in England. Under the impact of the metal, the lenses crack as does the safety glass used in automobiles, but the broken fragments will not scatter. The glasses are designed for athletes, workers, and others who run the risk of having their spectacles struck by flying objects.

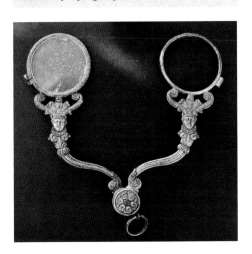

HEARING AID

Cupping one's hand around the ear allegedly increases hearing ability by five per cent—a minimal amount, but the shape created by the hand and the principles behind the action provided the basis for the production of ear trumpets and horns as early as the seventeenth century. Trumpets and horns placed to the ear would funnel the sound into the ear and amplify it for the hard of hearing. However, these protruding instruments would draw unwanted attention to the user; some people even went to the lengths of buying an 'under-beard' or an 'in-a-top hat' listening trumpet to disguise their use of the device. In 1819, King John VI of Portugal had a huge hearing aid built into a throne. People who wished to have a conversation with the King had to do so by speaking into the open mouths of two lion shaped resonators situated at the ends of the two arms of the King's seat; these conveyed sound to an earpiece near the head. Although these devices went some way to aid the hard-of hearing, it was not until the advent of electrical microphones and amplification that the terminally deaf truly benefited.

Alexander Graham Bell's experimentation in the 1870s that led to the invention of the telephone also provided the technology that would begin development of a portable hearing aid. The carbon microphone used in the telephone turned sound waves into electrical impulses and then converted them back into sound waves at the point of the receiver. Although American inventor Miller Reese Hutchinson designed and sold his "Acousticon" in 1902, the Marconi Company of England first utilised this technology to good practical effect. Marconi's "Octophone" used a carbon microphone and a valve amplifier to transmit sound to the wearer; its efficiency was somewhat detracted from by its weight, as when batteries were included the device weighed around seven kilograms. As technology improved with batteries and other components becoming smaller and lighter A Edwin Stevens produced a hearing aid that weighed only one kilogram and could be worn around the neck in 1935. Significant improvements in design occurred during the 1950s after the invention of the transistor, Microtone introduced its powerful and compact hearing aid using a transistor in 1953.

Hearing aids have evolved into three main types: those that fit into the ear, those that reside in the ear canal (both these types operate through air-conduction), and those that fit behind the ear, transmitting sound waves through bone-conduction. The digital age has brought with it much advancement and refinements, not least modern day hearing aids' ability to tune in and amplify to specific, selected frequencies, providing better sound quality.

1816
René Theophile Hyacithe Laennec

STETHOSCOPE

The stethoscope was invented by young French doctor René Theophile Hyacithe Laennec in 1816. The hollow, wooden cylinders, 30 cm in length, acted as resonating chambers and allowed physicians to listen to heartbeats and the function of lungs and stomachs; the instrument was named the "stethoscope" by Laennec, stethos being the Greek word for "chest".

It was in fact René's physical inability to put his ear to a certain patient's chest that gave him the idea of the stethoscope. His patient had a diseased heart but because she was obese Laennec could not gain information by tapping her chest, and contemporary etiquette meant he couldn't place his head on her body. He explained: "I recollected a fact in acoustics that augmented sound conveyed through solid bodies. I rolled a quire of paper into a cylinder and applied one end to the heart and one end to the ear and thereby perceived the action of the heart more clear and distinct."

Over time the invention became refined with the addition of two hearing devices which evolved into the modern-day binaural stethoscope, with bell-shaped chamber connected to the ear-pieces via Y-shaped tubing. The bell shape is naturally able to pick up low frequency sounds but it also has the addition of a diaphragm that allows it to pick up high frequencies as well. The stethoscope is a vital tool for all those who work within the field of medicine and has become a ubiquitous symbol of the doctor's profession.

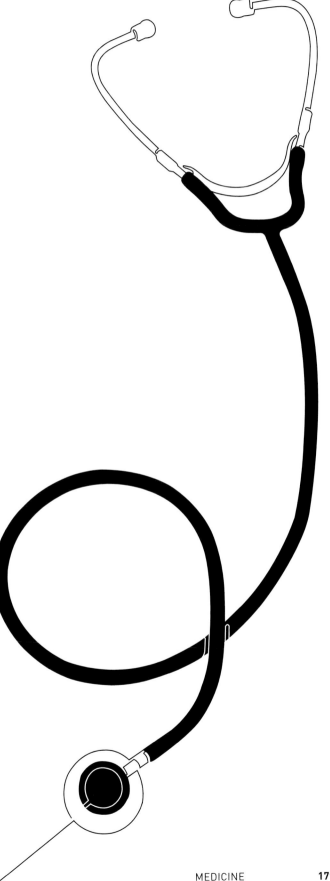

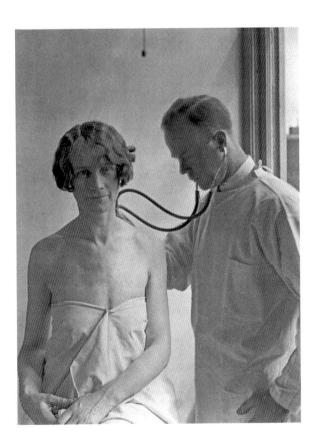

SCANNING

Top to bottom Ultrasound scan; x-ray;
Detecting Cancerous Tissues In Body Cells patent
filed 17 March 1972 by Raymond Damadian,
New York.

Wilhelm Conrad Röntgen was a German professor of physics who, whilst experimenting with a cathode-ray tube and fluorescent screen to research the effects of passing electricity through these gas-filled bottles in 1895, accidentally stumbled upon what he termed "X" rays. He had discovered a new way to produce radiation so that it was possible to view human bones through the flesh surrounding them. The first image he produced of his wife's hand with her wedding ring on is incredibly iconic.

During the first half of the twentieth century, X-rays were used prolifically in medicine, at least until it became clear that excessive use had potentially cancer-causing side effects. However, it was also ascertained that these rays could be used positively in the treatment of cancers through the processes of radiotherapy. Röntgen was awarded the first ever Nobel Prize for physics in 1901 and his discovery became the catalyst that sparked a revolution in the medical profession and gave birth to other forms of scanning such as computerised axial tomography, or the CAT scan, developed by British electrical engineer Godfrey Hounsfield of EMI in the 1960s; CAT scans used very focused X-rays that were taken in slices through the body allowing a 3-D image of a specific organ or bone to be created.

Obviously with the now heightened awareness regarding the damaging effects of X-ray technology, pregnant women were ill advised to be X-rayed, and so a new scanning technique would have to be developed. It was during the late 1950s that a Professor of midwifery at Glasgow University, Ian McDonald, made the connection between the wartime sound-echo device asdic, and the ability to scan a pregnant woman: his idea was based around the fact that if sonar could relate the position and shape of a submarine by showing an image of it on a screen, surely the same technology could be utilised in order to view a foetus. Computer enhancement turned echoes into visual images, and for over half a century parents have been able to see their unborn children via the use of ultrasound.

In 1971, America scientist Dr Raymond Damadian showed that MRI, or nuclear magnetic resonance imaging, could be used to detect tumours. By 1977 he and his team had created the first 3-D image of an entire human body by beaming high-frequency radio waves into a patient in the strong magnetic field of a whole-body scanner. Basically, a powerful electromagnet creates a temporary change in the constituent atoms in the body, the changes are analysed and used to generate three-dimensional images of internal organs. This technique's advantages lie in the way of producing no radiation whatsoever (unlike CAT scans) in order to produce these images. The technology is extremely sophisticated and very adept to recording biochemical processes making it perfect for monitoring muscular dystrophy and operations involving transplants.

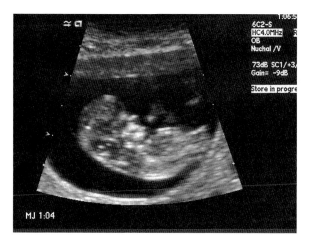

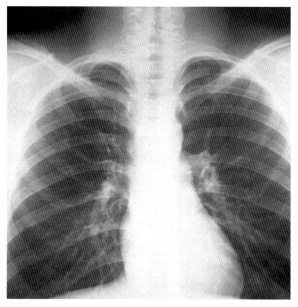

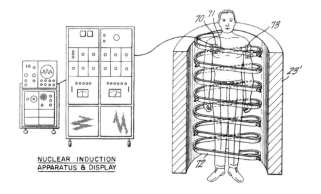

PATENTED FEB 5 1974   3,789,832

SHEET 2 OF 2

NUCLEAR INDUCTION
APPARATUS & DISPLAY

Sir Henry Gauvain

IRON LUNG

The iron lung was invented by Sir Henry Gauvain and developed into its first recognisable version in 1927 by Philip Drinker and Louis Agassiz Shaw of The Harvard School of Public Health. Its purpose was to maintain respiration artificially until an individual could breathe independently. Essentially, the patient was placed in an airtight metal box or cylinder powered by an electric motor with two vacuum cleaners changing the pressure inside the capsule and therefore pulling air both in and out of the lungs, imitating the body's natural physiological process.

The design of the iron lung was refined and the production cost cut in half by the inventor John Emerson, and was popularised in the United States in the 1940s and 50s during the height of the polio epidemic (a disease that attacks the nerves and can cause paralysis of motor functions and respiratory systems in a matter of hours, and is caught through ingesting contaminated food or water). Entire hospital wards were filled with the machines. During this time Drinker even provided instructions for building a makeshift emergency respirator for cases of life-threatening paralysis made out of wood and easily accessible hardware store materials such as car inner tubes and leather soles; the sufferer was to stay in this until a hospital iron lung became available.

Polio victims usually had to be kept in the iron lung for anywhere from a week to three week, with mirror angled above the head affording a skewed and unique perspective of the world as backward and upside down, as being "incomplete embryos in a metal womb".

With the success of worldwide polio vaccinations there are hardly any new cases of polio. However, there are still estimated to be around 40 polio survivors still living in the iron lung.

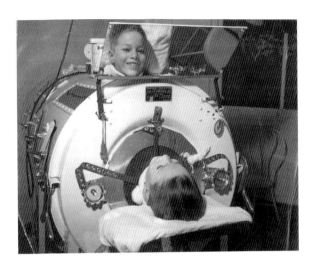

Dr Willem Kolff

ARTIFICIAL HEART

Top Patent for an artificial heart, filed by James A Tindal, California, 23 December 1969.

Bottom CardioWest™ temporary total artificial heart.

Essentially a muscle that acts like a pump, the heart is functionally very simple. Despite this, it is also extremely intricate, and ultra-sensitive to the subtle changes in the body; subtleties that are almost impossible to replicate artificially. This is why the design of a fully functional synthetic heart replacement is regarded as the holy grail of modern medicine.

From the 1950s onwards, mechanical hearts have been used successfully to replace the heart during surgical operations, but only for a relatively short period of time. The first patients to receive artificial hearts proper were animals. Dr Willem Kolff dedicated much of his life to the development and invention of artificial hearts. He implanted an example into a dog at Cleveland Clinic, keeping the dog alive for 90 minutes. Kolff established the Division of Artificial Organs at the University of Utah in 1967. A series of experiments with artificial hearts and calves followed, culminating in 1981 when a calf called Alfred Lord Tennyson survived for 268 days via the use of a Jarvik 5 artificial heart, so named after a graduate student working for Kolff.

Kolff is the name most commonly connected with this very specific field of medicine. Within his first patent, Jarvik explains the many factors that need to be taken into consideration when designing an artificial heart, including avoiding foreign-body rejection, preventing excess heat generation, and keeping the noise produced to a minimum, as well as not restricting the patients mobility. The design relied upon pumps powered by an external power source; it works by alternating contractions and relaxation. Located at the bottom of the heart is a pump in a sac of water, which mimics the organ's functions with the aid of electronics, the sac expanding and contracting to force blood in and out of the heart.

The first artificial heart to replace a human heart was the Jarvik-7 model in 1982; the patient Dr Barney Clark was on the verge of death when he received the device, and survived for only 112 days after the operation. Four more patients suffered a similar fate until the experiments were stopped. However, Jarvik persevered and as technology improved so did his artificial organs. The Jarvik-2000 model has been in recent use; the actual machine is the size of a thumb and has a very thin power line to a transmitter the size of a small coin, which is screwed into the skull behind the ear. A battery pack is worn around the waist. These artificial hearts cost $50,000 to install—a small price to pay considering how much money it costs health services to provide care for patients with heart disease.

One of the world's leading heart transplant specialists, Frenchman Alain F Carpentier, announced in October 2008 that a fully implantable artificial heart would be ready for clinical trial by 2011, and for full transplant in 2013. The prototype uses electronic sensors and is made from what are known as "biomaterials" or "pseudo-skin"—basically chemically treated animal tissues.

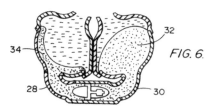

U.S. Patent Nov. 13, 1979 4,173,796

FIG. 6.

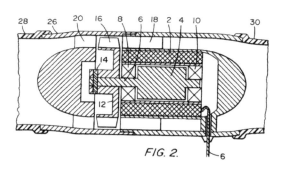

FIG. 2.

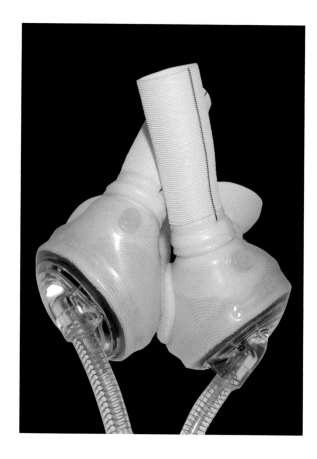

PACEMAKER

Bottom Patent for a pacemaker that requires no plug-receptacle or electrode catheter, filed 26 June 1991 by Hector O Trabucco and Jordan Gavrielides, Argentina.

The pacemaker regulates the beat of the heart via the use of external electronic impulses; the heart regulates itself in response to its own electrical impulses, but sometimes they can become irregular and cause various types of ailments and diseases, sometimes with fatal consequences.

Paul Zoll at Beth Israel Hospital, USA, invented the first external shock-maker, testing it on a 65 year old man with congestive heart failure on 7 October 1952. A needle was inserted into the apex of the heart with another electrode placed upon the skin surface on the ribcage. Once the patient's heart had stopped, Zoll used his device to keep him alive by delivering small electrical shocks for 52 hours. However, the process was painful and resulted in burns if the patient was connected to the equipment for over a day. On 9 October the man's heart began working by itself again and the patient was discharged from hospital.

Improvements occurred very shortly after; Clarence Lillehei of Minneapolis University was convinced that electrodes could be attached directly to the heart as long as the electrical current was rigidly controlled. He sought the technical help of former TV repairman Earl Bakken, who devised an external power source the size of a small cigarette packet. The technology was progressive but the small open hole in the chest necessary for the insertion of wires was unhygienic and frequently became infected. Lillehei reflected that the next step was to "make it a little smaller, coat it with plastic, and put it inside the body... the first implantable pacemaker". This was made possible by the invention of the transistor, which allowed the device to be smaller than previously possible. In 1957, Earl Bakken set up Medtronics, a company that specialised in the making of pacemakers; two decades later it was making $180 million a year.

Although pacemakers were almost an overnight success in the medical world, their long-term success was still under inspection. If stimulated at a certain point during a single heartbeat the heart could flutter. If this happened repeatedly, as was possible with a pacemaker, it would kill the patient within a few months. The problem was solved by the "demand pacemaker", invented by Barouh Berkovits in the late 1960s. Berkovits' pacemaker has extended hundreds of thousand of lives by, simply detecting when the heart does not beat and electrically stimulating a beat.

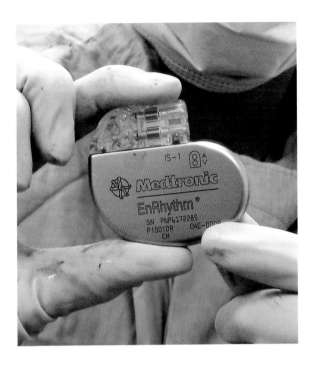

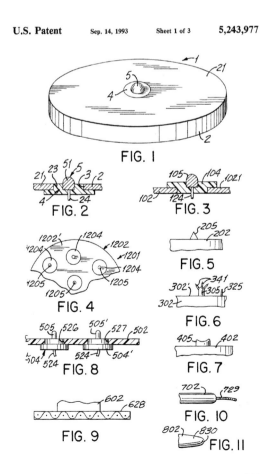

PROSTHETICS

The earliest artificial limb dates from around 300 BC and was unearthed from a tomb in Capua, Italy in 1958. It was a simple copper and wood leg. Some of the earliest accounts of prosthetic limb use were recorded in Ancient Greek and Roman times, including one concerning the roman general, Marcus Sergius, who lost his right hand in battle during the Punic War; in order to return to battle he had to have a replacement hand fashioned out of iron so that he was able to carry his shield.

The development and evolution of artificial limbs is inextricably linked with war and the injuries sustained through periods of war. During the fifteenth and sixteenth century craftsmen who made military armour also designed and made iron prostheses, and a major contribution for the advancement of prosthesis was made by French military doctor Ambroise Pare. In 1559 Pare devised safer and more effective methods of amputation as a lifesaving method, meaning a much higher proportion of amputees survive, therefore heightening a requirement for post-op prosthetic limbs. He turned his attention to designing and making artificial limbs in a more scientific manner, striving to simulate a degree of natural movement. Pare invented a hinged hand with fingers that worked individually by the means of cogs and levers, prosthetic legs featuring movable and locking knee joints controlled by a string, and a flexible foot fitted with a spring, all with specialised attachment harnesses.

Other notable advancements made in the development of artificial limbs up to the modern day include Dutch surgeon Pieter Verduyn's development of a lower-leg prosthesis with specialised hinges and leather cuff improving attachment in 1696, and in 1863 Dubois Parmlee of New York City developed a fastening body socket to the limb using atmospheric pressure. Then in 1898, Dr Guiliano Vanghette invented an artificial limb that moved through muscle contraction.

Today's artificial limbs are constructed from plastics and foam, making them immeasurably lighter than their iron prototypes and has also made it possible for cosmetic modelling to match a patient's skin colour and body shape. With the introduction of computer-aid design programs the fit of sockets and artificial limbs has vastly been improved along with the addition of microprocessors and sensors being included in prostheses to improve stability and gait.

More recently we have seen the progress of sports prostheses. South African Sprinter and double below-the-knee amputee, Oscar Pistorius, was told he was unable to participate in able-bodied competitions because his of his "Cheetah" prosthetic limbs; track and field's governing body ruling that it provided him with an unfair advantage. The J-shaped "Cheetah" limb, so called because of its striking resemblance to a Cheetah's hind-legs was designed by American inventor and amputee, Van Phillips, and was introduced during the 1980s and updated with patents through the 1990s. It is made out of carbon-fibre based material, and works by applying weight when a runner lands on his heel, which in turn is converted into energy and literally springs the athlete forward. In Phillips' own words: "The Cheetah may be more advantageous than the human foot."

FALSE TEETH

False teeth or dentures were fairly uncommon until the eighteenth century, when the use of sugar in food became more widespread. Teeth made from bone, ivory, or even cloth were commonplace, tied to the remaining teeth with silk threads. 1770 saw the first porcelain dentures made by a Mr Alexis Duchâteau, with the first British patent filed in 1791 by Nicholas Dubois de Chemant. It was not until the mid-nineteeth century, with the invention of vulcanised rubber, however, that false teeth became widespread, as they could be moulded to the wearers remaining teeth and gums; with human teeth being widely used until the use of acrylic resin and plastic in the twentieth century.

Top left Denture patent with a bimetallic strip that fits to the gums by responding to body heat, filed 22 April, 1971 by Douglas W Powell.

Bottom right False teeth made from hippopotamus tusk, dating from circa 1854.

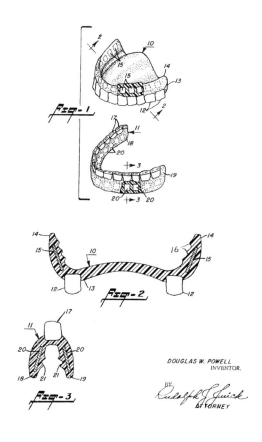

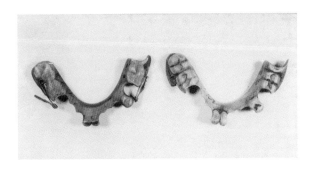

1991
Pfizer Pharmaceuticals/
Andrew Bell, David Brown, Nicholas Terret

Middle Chemical composition of Sildenafil Citrate, or Viagra.

Bottom Natural Viagra.

VIAGRA

Viagra, the small, blue, diamond-shaped pill—now popularly regarded as the first artificial aphrodisiac—originally started life as a drug created to treat high blood pressure and cardiovascular diseases, and was developed by Pfizer Pharmaceuticals. The first patent for the pill was awarded to Andrew Bell, David Brown and Nicholas Terrett, in 1991. The compound pill, then known as sildenfal, was tested on sufferers of angina, and whilst having insufficient effect on their heart conditions, it unexpectedly gave certain patients the ability to maintain an erection where this had previously been a problem. The compound worked by blocking an enzyme called phosphodiestrase that destroys cyclic GMP, a molecule that makes blood vessels relax, thus boosting the blood-flow to the penis. Terrett and fellow Pfizer colleague, biologist Peter Ellis, began research into how the drug could be utilised for erectile dysfunction, and filed a patent in 1994 for the pill to be used solely this reason.

After Pfizer had carried out another three years of research into using sildenfal to enhance penile erection rather than for treating cardiovascular problems, chemical engineers Peter Dunn and Albert Wood devised a way to mass produce the compound through a nine-step process, essentially creating the pill available today. Though Pfizer representatives insist that the development of the drug was wholly a team effort, Terrett has since self-assigned credit to its discovery, stating that "me and my team discovered how useful the drug might be... they (Wood and Dunn) created a way of mass producing it only."

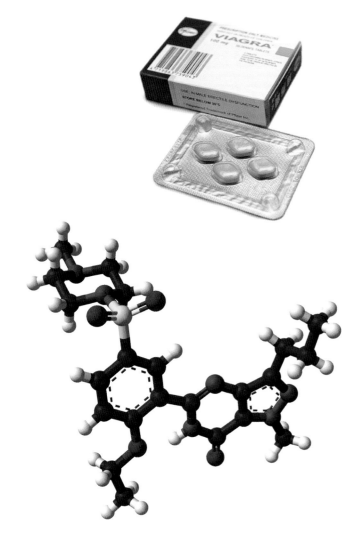

circa 1640

CONTRACEPTION

Apparently named after the court doctor who was given the task of explaining to King Charles II what the device was used for, the earliest historical remains of a condom date back to 1640, and were found in the foundations of Dudley Castle, England. However, the use of rudimentary condoms has been occurring for some 3,000 years: Ancient Egyptian art depicts the use of animal gut as a barrier method of contraception.

The use of rudimentary condoms is therefore ancient but the principle reason for using one has changed significantly over time. Gabriel Fallopius, a sixteenth century gynaecologist, advocated the use of linen sheaths that were doused in solution then left to dry before use, before being tied to the penis with a ribbon. Fallopius claimed that the condoms protected against syphilis or venereal diseases. Contemporary thought stated that in a male-oriented world, the onus was on avoiding sexually transmitted diseases, not unwanted pregnancy. That the sheaths may have served this dual purpose would not have been a reason for widespread acclaim at the time.

Animal intestine condoms were still available up until the 1960s, but it was the advent of vulcanised rubber, developed by Charles Goodyear in 1843, that allowed condoms to be made and sold cheaply, whilst at the same time increasing their reliability. The latex condom was the most popular form of contraception up until the Pill, created by chemists Gregory Pincus and Min-Cheu Chang, was introduced to the world in 1960. Devised from previous studies into the wild yams of Mexico that produced the female hormone progesterone, the pair realised that a modified version could be used to prevent conception. Backed by feminist nurse and birth control propagator Margaret Sanger—who raised $115,000 for research purposes—Pincus produced Norethynodrel, which gave women control over reproduction, and coincided with the women's rights movement. However, the Pill isn't without its problems, often causing unwanted side affects that see many women still opt for using condoms so as to not unsettle the hormonal balance of their bodies. Furthermore, although pregnancy rates decreased, the amount of young people contracting STI's increased as use of the Pill saw less individuals wearing condoms.

The 1980s saw progression in contraceptive technology with the development of the PC4 (post-coital)—or the 'morning after pill'—that could be taken within three days of unprotected intercourse and prevent conception, and also the introduction of the abortive RU486 pill in the 1990s. During this proliferation of pill- based contraceptives the condom had gone full circle in regards to its principle purpose, being used in a scheme to prevent the spread of AIDS throughout Africa.

Other methods of male-concentrated contraception are being developed. Scientists at King's College London are working on a birth control pill that stops a man ejaculating, apparently with no depletion to the man's personal satisfaction. The pill only has to be taken immediately prior to sexual intercourse, and effects wear off after 24 hours. The idea first appeared when scientists noticed that drugs used to treat high blood pressure and schizophrenia also had the unforeseeable side effect of preventing ejaculation.

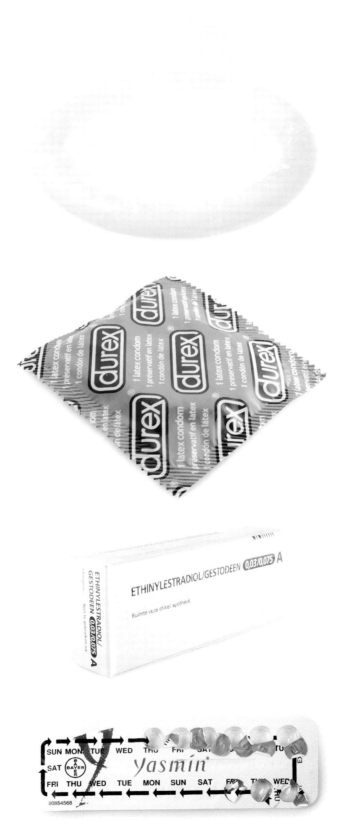

1965
Robert Edwards

IN VITRO FERTILISATION (IVF)

Robert Edwards of Cambridge University sowed the seeds in creating the process known as in vitro, or in glass, fertilisation. Using previous research he had carried out on different animal species as his basis, Edwards believed that the work could be extended in order to treat women who had blocked fallopian tubes, meaning that eggs could not travel to the womb in order to be fertilised by sperm. The process would involve the scientific fertilisation of one or more embryos outside of the body, which were then replaced back into the womb using a microneedle and by 1965, with the help of two gynaecologists Howard and Georgeanna Jones, Edwards had successfully adapted the techniques to befit human patients.

Over the next few years of research Edwards encountered many problems and flaws that resulted in fruitless tests. This was until young student Barry Bavister supplied a culture medium that proved successful; however another problem lurked—it quickly became apparent that the immature eggs that Edwards was extracting from the ovaries to cultivate in the culture medium did not develop for very long. However, in Oldham, gynaecologist Patrick Steptoe frequently encountered ripe eggs whilst performing laparoscopy (key-hole abdominal surgery). Edwards and Steptoe joined forces and started to give patients small doses of hormones that would facilitate the production of multiple ripe eggs, and also calculated the best times at which to harvest these eggs. Unfortunately, the first baby that was conceived via the breakthrough work of these two IVF pioneers in 1975 had to be terminated due to its mother, Marlene Platt suffering an ectopic pregnancy.

The pair decided against the use of hormone-induced egg-production and became fully reliant upon the natural menstrual cycle. It was Lesley Brown who underwent IVF treatment in 1977; just eight cells were transferred into the hopeful patient, and on 25 July the following year Louise Brown was delivered successfully by Caesarean section.

Although the technology is an incredible demonstration of the power of science, and the conception and birth of Louise Brown through IVF treatment came to symbolise this progression, the process is still only successful in 23 per cent of cases.

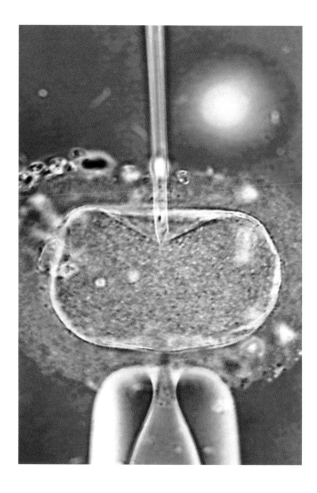

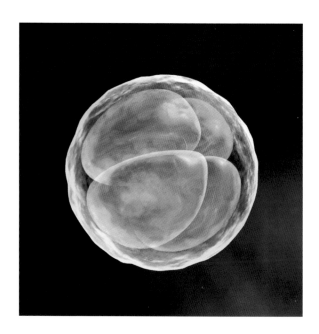

1897
Felix Hoffman

ASPIRIN

Aspirin, or acetylsalicylic acid, has its origins in natural remedies. Its active agent, salicylic acid, was first extracted from the plant Spiræa; it was observed by the reverend Edmund Stone in 1763 that the farmers of Chipping Norton would use white willow tree bark to cool fevers and alleviate aches and pains. 60 years later, Italian chemist Raffæle Piria revealed that bark of the white Willow contained salicylic acid.

However, salicylic acid had severe side effects that included bleeding in the stomach and even death. When German company Bayer wanted to utilise the acid, a solution to the inadvertent problems caused by the substance had to be devised. In 1897, Bayer employee Felix Hoffman rediscovered an abandoned theory—devised in 1853 by French chemist Charles Frederic Gerhardt—using an acetyl compound of the acid was the most effective solution, after testing several different examples on his arthritic father. Renamed as aspirin, the now ubiquitous painkiller was born.

Aspirin inhibits the production of blood clotting hormones and the hormones that make nerve endings sensitive to pain. It can also protect against nerve damage and protect against bowel and colon cancer. The drug thins the blood, therefore reducing the risk of both heart disease and strokes, and can block the enzymes that cause inflammation, pain and fevers. More recently, compounds have been developed to counteract the imbalance between the drugs affecting both these enzymes, and those that are essential to maintaining health in the kidneys and liver.

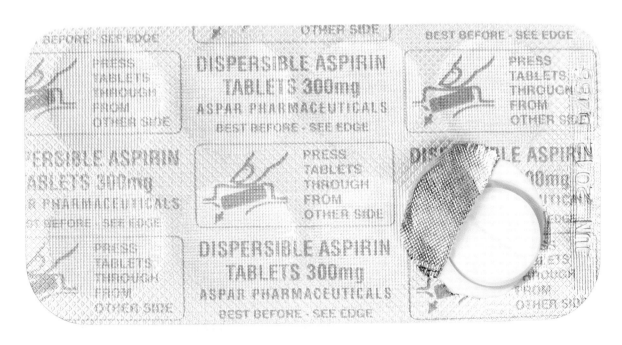

GENETIC ENGINEERING

Breakthroughs in genetic manipulation—also known as genetic engineering—began in 1956 when Arthur Kornberg purified DNA polymerase I from E. Coli, the first enzyme that allowed DNA to develop in a test tube. This development was shortly followed by the groundbreaking 'cracking' of the genetic code by a team led by Marshall Nirenberg and H Gobind Khorana in 1966. However, it was not until 1977, two years after Fred Sanger developed the chain termination method for sequencing DNA, that the first genetic engineering company was founded.

The company, Genentech, used recombinant DNA methods to make medically important drugs for the first time, changing the face of modern medicine and entering medical science into a new era of controversy over ethical practices and even the values of human and animal life.

Cancer genes were identified in 1981, the chromosome responsible for Huntington's disease in 1983, and by 1988 a project was in place to identify the entire DNA sequence in human chromosomes.

The Human Genome Project, as it was known, remains one of the largest single investigational projects in modern science, and by 2003 had achieved its aim, although research continues. The implications of the project have already been vast, simultaneously stimulating major medical breakthroughs and fevered ethical debate. As with most major breakthroughs that challenge the equilibrium of human awareness, the implications of genetic engineering and the Human Genome Project caused reverberations through governments and religious bodies, no more visibly than in the United States of America, where the government had originally funded the project through the National Institute of Health. In August 2001, Former president George Bush turned on progress in genetic engineering by issuing a presidential decree banning the use of federal funds for research on new human embryonic stem cell lines. The backlash had a suffocating financial impact, whilst also placating the Christian Right. Dr Kevin Eggan, who had been researching into human embryonic stem cell generation, found himself isolated overnight, but thanks to the generosity of Harvard where he carried out his research, he was able to carry on. In the UK, the Wellcome Trust also played a valuable role in funding the Sager Centre in its research.

The concerns surrounding genetic engineering revolve around the issue of recreating life, and whether or not it is the responsibility of humans to do so. The famous creation of Dolly the sheep, the first cloned animal, was for many the very articulation of that fear, and subsequent animals developed specifically for meat are a continuation of it. However, the potential of genetic engineering in medicine is vast.

It is the ability to regrow skin and organs that has most excited the medical and scientific community. Already adult stem cells are used in, for example, bone marrow transplants for leukaemia, and if the power of stem cells to grow new tissues can be harnessed, doctors may be able to treat diseases such as Alzheimer's by using stem cells to replace missing structures in the brain.Developments are already suggesting that the regrowth of organs and limbs may soon be possible.

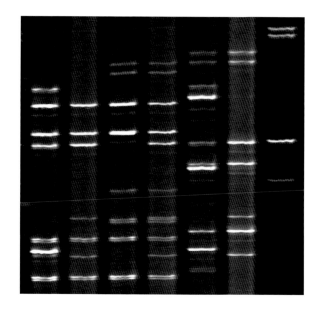

1996
Ian Wilmut and Keith Campbell

CLONING

The invention of the process that produced Dolly the sheep on 5 July 1996 became surrounded by controversy when an official announcement was made in February the next year. The process was developed by geneticist Ian Wilmut and cell cycle biologist Keith Campbell for the Roslin Institute in Scotland, and involved taking the cells from an adult sheep, culturing them in vitro, and transferring the data to an enucleated egg which was then implanted into a surrogate mother. This created offspring that were identical to their mother, like twins. Dolly was cloned from tissue taken from the udder of a six-year-old ewe; an impressive feat due to the prior belief that cloning was only possible if cells were extracted from early embryos, rather than adult animals.

Some viewed the new technology as a positive technological step; farmers welcomed the ability to uniformly breed top-quality livestock and the ability to improve their herds; animal organs could be cloned alleviating the demand for donors and the waiting times of patients; and parents who feared that they could pass a genetic effect on to their unborn children could have a fertilised ovum cloned and the duplicate tested, so that if the clone were free from defects then the other could be implanted. Opponents believed that the process was completely unnatural and therefore unethical, the notion of scientists effectively playing God unnerving many members of religious groups and the political Right. Furthermore, some critics of the cloning process even envisaged an inevitability, so far unfounded, of human parts 'farms'.

The first commercial use of this technology by the Roslin Institute was in the creation of Polly the sheep which had the human gene so that her milk included human factor IX, a blood-clotting agent needed by individuals with haemophilia. There is a general consensus that experiments like Polly and 'therapeutic cloning'—the production of tissues and organs—although steeped in controversy, could aid medicine and science in amazing and vast ways. However, it has been globally agreed that so-called "reproductive cloning" of humans should be banned.

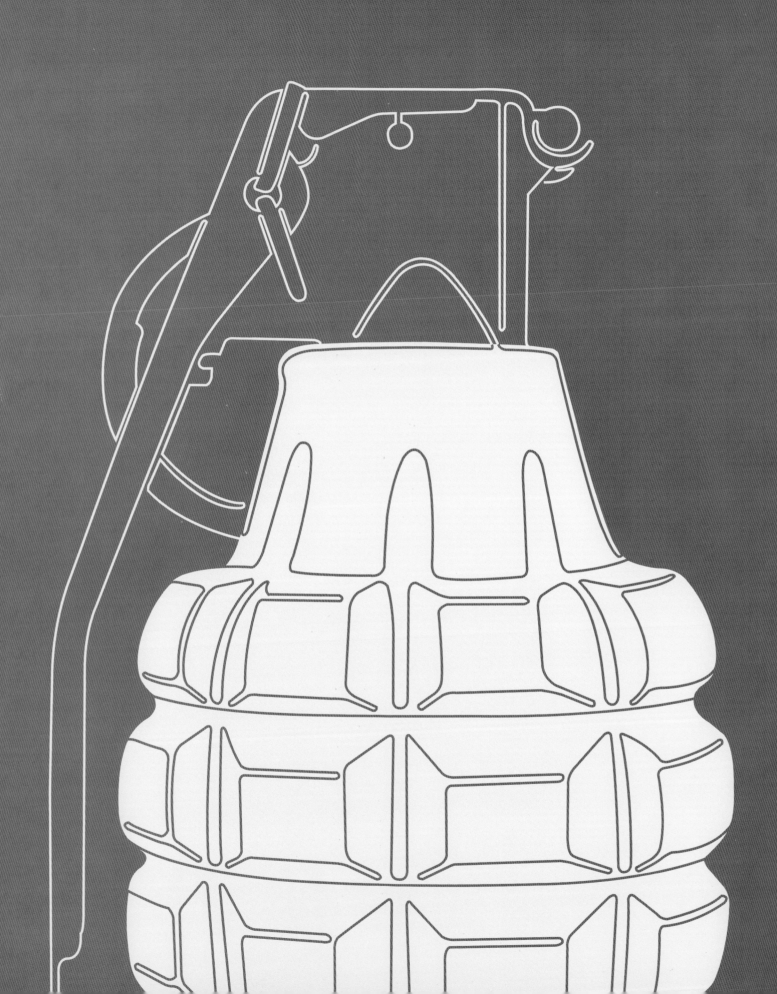

warfare

BOWS OF WAR

The earliest representations of Chinese crossbows date from 400 BC when the Chinese army had 50,000 crossbowmen equipped with mass-produced weapons. These crossbows—discovered in tombs in China dating back to the fifth century BC—ornately carved in bone and wood, were an important weapon capable of firing up to 650 feet. The first pictorial evidence of European hand-held wooden crossbows—dating to around 1200 AD—were introduced from the East during the Crusades. Religious leaders who considered them to be inappropriate for Christian warfare initially outlawed these crossbows; however, their effectiveness in battle soon prevailed over any theological concerns.

Early bows would be mounted on a headstock, and the string drawn back via the use of either a lever, or later, a rack-and-pinion winding mechanism. The Chinese designed a modified version that went some way to solving the crossbow's main drawback—that of vulnerability from attack. Essentially they built the first machine crossbow, which had around ten 'bolts' and corresponding drawstrings that could be shot in sequence one after the other.

Between the twelfth and sixteenth century, popularity for the weapon began to dwindle. The crossbow's major design flaw was its short range in comparison with the longbow—demonstrated in 1346 at the Battle of Crecy. Here, Philip VI of France took on the weaker forces of Edward III of England and lost due to being outranged by the English longbow men. Originating in the Welsh borderlands, the longbow could shoot an arrow a metre in length a minimum of 200 metres and pierce the armour of the enemy with ease. The longbow was used to great effect during the Hundred Years War against the French, creating a superstitious image of the English archer and his formidable bow.

Popularity for the weapon dwindled with the advent of gunpowder and today the crossbow is used solely for sporting activities.

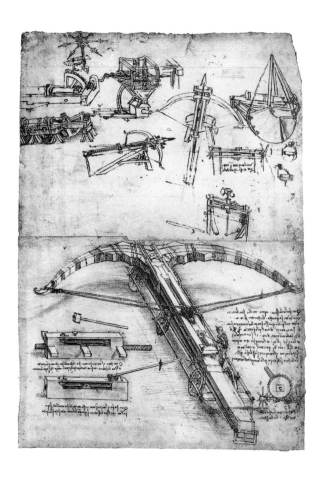

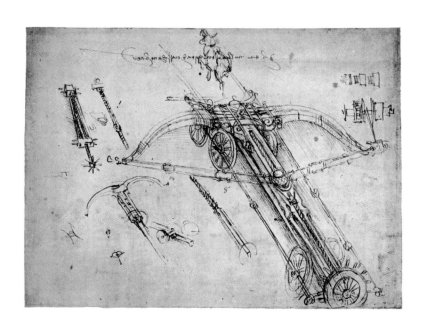

Top and bottom Leonardo da Vinci's designs for conventional and oversized crossbows.

**circa 399 BC
Ancient Greece**

CATAPULT

Though an antiquated instrument with a disparate history across the globe, the catapult's recorded history began in Greece in around 399 BC, when the historian Diodorus Siculus first recorded the use of a device that mechanically fired arrows in the manner of an enlarged crossbow. More concrete practical implementation was recorded soon after, against the Carthaginian city of Motya in 397 BC, during the second Sicilian War. As the century continued, the Greek focus on developing arrow-firing projectile weapons increased. A description of the contents of the Athenian arsenal during the fourth century includes a list of catapults armed with a variety of projectiles in different shapes and sizes.

Roman development saw the invention of the arcuballista, a giant projectile machine resembling a flexible crossbow, notably employed by Alexander the Great in siege situations. As weapons technology developed, designs that relied on torsion to propel their load gradually became commonplace; they delivered far more power and covered a greater distance than the previous models, although given the natural arc of the projectile, they were less accurate.

The instrument continued to be used as a siege weapon long into the Middle Ages. Designs included the crossbow-resembling *ballista*, and its derivative springald; the mangonel, a one armed design which launched projectiles from a scooped 'bucket' at one end; and the *trebuchet*, a large design which utilised a sling to fire heavy loads of stones hundreds of meteres. As a unique point in securing the significance of the catapult in the Middle Ages, the initial spread of the Black Death through Europe was partially the result of the weapon's use in siege warfare; the plague spread to Western Europe with Genoese merchants, infected after an Asian Tartar force launched infected bodies into the fortified Crimean town of Kaffa, with the subsequent fleeing dignitaries and tradesmen carrying the disease by boat as they went.

By the fourteenth century, the invention of the cannon—with its mechanical and speed advantages—quickly took over in military situations. Nevertheless, France did give catapults another outing much later, during the First World War, using them to fling grenades beyond enemy lines.

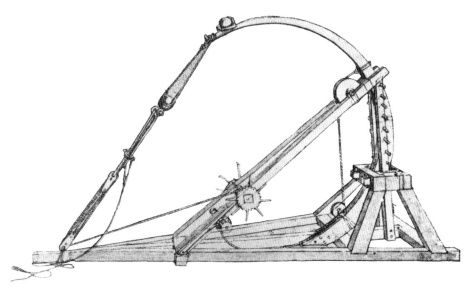

Right A design by Leonardo da Vinci for a catapult.

HAND GUNS

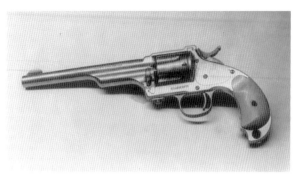

Early fourteenth century handguns were nothing more than small cannons. Gunpowder was poured down an iron tube that acted as the barrel, closely followed by a lead ball. A match was lit at the touchhole that set light to the main charge in the barrel, creating a huge accumulation of gas, which led to the forcing out of the lead ball at a great pace.

By the late fifteenth century, the matchlock musket had been invented which had a lengthened barrel and wooden stock, and an evolved mechanised ignition system that plunged a lit match into the flash pan to ignite the primer.

The sixteenth century saw musketeers replace archers in battle. Knights' armour was not strong enough to stop bullets and new designs made to deal with this were far too heavy and cumbersome to fight in. Eventually, most infantry and cavalry did away with armour almost completely, as it served no purpose in wars that no longer involved hand-to-hand combat.

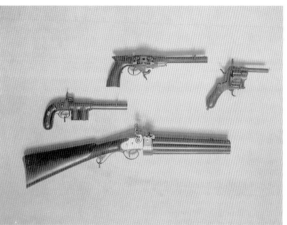

The matchlock system of ignition was susceptible to wet and windy weather conditions, meaning that the lit match had to be constantly monitored. This led to the invention of the wheellock in the early 1700s. The wheellock involved a serrated wheel which was set spinning when the trigger was pulled. It revolved on iron pyrites to create a shower of sparks that ignited the priming powder. This system allowed for the development of pistols that could be loaded and carried about one's person, ready to be fired at any time. However, although more reliable, the wheellock mechanism was still not as reliable as most marksmen would have desired, and in addition it was very expensive. Thus, the system was simplified to a piece of flint scraped down a steel plate—coined the "flintlock". It was cheap, reliable, and could be made to any size, making it the most popular form of ignition for firearms for over 250 years.

The system was simplified further still in 1805 when Scottish amateur chemist and gun enthusiast Alexander Forsyth patented a mechanism that relied upon chemicals generating sparks when struck. 20 years later, chemicals were placed in small copper 'percussion' caps that were unaffected by external elements and needed only a simple mechanism to trigger an explosion. This system opened up potential for a whole new range of multi-shot weapons, one of the first of which was Samuel Colt's revolver, made in 1836. However, the bullet and powder still had to be loaded into the gun's muzzle, but ideally everything would be loaded from the breech end. The Smith & Wesson Company of America were responsible for the solution to this problem; a metal-case cartridge that had the detonating chemical in the base. These bullets were then loaded into the cylinder of the revolver, exploding when the hammer struck the base. The next step was breech-loading metal cartridges that allowed the British infantryman of the First World War to fire ten rounds a minute from their Lee Enfield rifles, fast enough to give enemy German soldiers the impression they were facing machine guns. They became the basis of all modern hand-held guns, paving the way for the development of the self-loading pistols of the modern age.

Top to bottom Jesse James' 44 Hopkins & Allen pistol, 1873; a collection of revolvers, circa 1926; Berreta 92FS.

Opposite Smith & Wesson model 686 revolver with cylinder open.

1861
Richard Gatling

MACHINE GUN

Though the development of handheld revolver pistols went some way to the desired ideal of being able to fire multiple shots without reloading, the advent of machine gun design was the point at which the idea really took hold.

The Gatling gun, invented by American doctor Richard Gatling in 1861, utilised ten barrels in a cylindrical form. Cartridges from a magazine were dropped into the breeches of each individual barrel via rotation caused by the turning of a handle. The gun could fire 3,000 rounds a minute and was adopted by many armed forces around the world.

The first fully automatic single barrel machine gun was invented by an American working in London. Hiram Maxim was advised by an American friend that the route to riches was to "invent something that will enable these Europeans to cut each other's throats with greater facility". By 1884, he had created a machine gun that was fed by cartridges that were loaded onto a belt; the mechanism pulled a cartridge free, loaded it into the breech and then fired it. It then ejected the empty case and used the energy of its recoil action to reload and repeat the sequence, which would be continuously repeated for as long as the trigger was held down and cartridges were available.

Many armies took up the gun invented by Maxim and it was responsible for massive destruction within the First World War; British forces alone suffered 60,000 casualties in just one day at the Battle of the Somme in July 1916. They were used as weapons of defence and essentially stopped the mass-attacks on foot over open ground as it left the approaching forces dangerously exposed, turning the First World War into a static battle where armies were left immobilised in trenches.

Top Etching of a Gatling gun.

Left American soldiers with an early machine gun.

Opposite A Browning machine gunner, USA, 1942.

1915
Landships Committee

TANKS

Top A 1917 propaganda poster for the tank.

Bottom left and right Detail and a model from a Leonardo da Vinci drawing of an early armoured vehicle.

Opposite top An American Patton tank.

Opposite bottom A modern-day Russian tank.

The tank developed from an aversion to the horrors of the Western Front. The British were looking for a vehicle that could cross trenches, destroy obstacles such as barbed wire, and keep on going in the face of machine gun fire. The Rolls-Royce Armoured Car, developed in 1914 and used by the Royal Navy Air Service in the First World War and early Second World War, was the only vehicle that came vaguely close to the desired specifications.

The Landships Committee, sponsored by the First Lord of the Admiralty Winston Churchill, took their lead from this humble Rolls Royce, and from it created the first successful prototype tank, known as "Little Willie". It was put to the test for the first time on 9 September 1915.

Originally termed a landship, the vehicle was referred to simply as a "water-carrier" in order to keep its actual purpose a secret from the workers that built it. Believing that they were building water tanks, the landships were nicknamed "tanks" by the workers, and the title remained.

The first Mark I tank entered service during the Battle of the Somme, rendering trench warfare obsolete. The tanks made a significant contribution to victory in the First World War, but the Germans responded by making trenches wider to trap vehicles, and developing weaponry to blow them up. They also introduced their own A7V tank—a monster of a machine that could withstand far more firepower than British and French designs.

Throughout the Second World War, tank technology advanced significantly. Germany deployed Panzers, which were much more nimble, eschewing heavy armour in place of greater speed and manoeuvrability. However, while this was a significant advantage, they remained vulnerable in direct combat. Nevertheless, the Panzer developed quickly—the Panzer IV, also known as "the Tiger", weighed under 20 tonnes, a significantly lighter specimen than the previous models.

The tank remains in regular use in combat environments, but it is still not the perfect vehicle of war. One of the greatest continuing limitations of the tank is visibility; despite design improvements, infantry still tend to be spotted mounted on top of the vehicle, entirely exposed. Meanwhile, for those submerged within the body of the tank, the problem of escape from the metal body in the case of problematic assaults is still prevalent, and responsible for numerous fatalities in modern war zones.

GO OVER THE TOP WITH THE TANK THAT WILL WIN THE WAR!

The watchword on the turret should inspire the crew that man the guns. Only by united action in America's war industries can we crush out disloyal and treasonable efforts to obstruct.

We must stand together, work harder, produce more and aim straight to

"KAN THE KAISER"

LAND MINES

A land mine is an explosive the activation of which is triggered by a weight or pressure. The name comes from the practice of mining in the field of war, where collapsible tunnels were dug under enemy fortifications in order to destroy the fortifications above. The mine is a vicious form of weaponry, entirely uncompromising when it comes to choosing its victim. It is also controversial because after conflict, a hidden land mine can lie dormant indefinitely, leaving vast areas of land either unusable or highly dangerous; this has been particularly reported upon in the late twentieth century around the countryside of Cambodia and Bosnia. It was in fact in Ancient Rome that the first mine, in its most rudimentary form, was used—the Romans dug small, covered holes that gave way when stepped on. Hidden within the hole was a sharpened spike, intended to mutilate the foot of the victim. In the Middle Ages, similar devices consisting of spikes were scattered across battlegrounds, to delay and injure oncoming enemies.

The origins of explosive devices, such as those seen today, can be traced back to 1277 AD, when the Song Dynasty Chinese used what they described as an "enormous bomb"—an exploding shrapnel device called a *tetsuhau*—against an assault of the Mongols, who were besieging a city in southern China.

The first modern high-explosive land mines were created by Brigadier General Gabriel J Rains during the Battle of Yorktown in 1862, and were known as "Rains' mines". Germany, around the start of the First World War, developed mine technology further, and they were copied and manufactured by all major participants during the war. The British even manufactured poison-gas land mines, a resurgence in the manufacture of which has allegedly been considered by both the Soviet Union and the United States at various points throughout the last decade.

The International Campaign to Ban Landmines successfully moved to prohibit the use of the weapon with the 1997 Land Mine Ban, with 158 nations signing a subsequent anti-land mine treaty. Notably, 37 countries declined to participate, including the United States, Russia, China, India, and Israel.

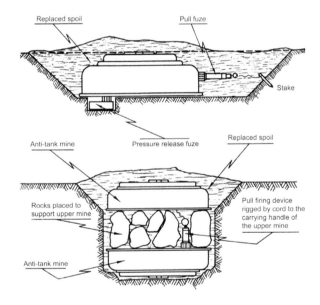

Top right Retrieved land mines in Siem Reap, Cambodia.

Bottom right A land mine implementation diagram.

Left An Italian anti-tank mine and a section through an anti-tank mine.

1891
Karl Elsener

VICTORINOX SWISS
ARMY KNIFE

The development of the Swiss army knife was directly the result of national pride. Karl Elsener, a patriotic Swiss surgical instrument designer, was outraged at the fact that the knives supplied to the Swiss Army were in fact German models, and resolved to do something about it using his professional knowledge and skills. The 1891 'soldier's knife' was the result; a wooden-handled design with integrated blade, screwdriver, can opener, and punch. Despite the fact that it was a success and was sold to the army, Elsener felt driven to improve it, adding a corkscrew, a second blade, and a spring mechanism. With all these complex parts loaded into one device, the Victorinox Swiss army knife as it is known today was born.

The distinctive cross logo was carried over from his surgical implements, but the famous name Victorinox, which became the title of his knife-distribution company, was a combination of his mother's name and 'inox'—the type of steel that he used to make the knives themselves.

During the Second World War, American GIs named Elsener's invention "the survival knife", an endorsement that cemented its reputation as an invaluable wartime tool.

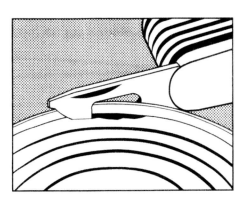

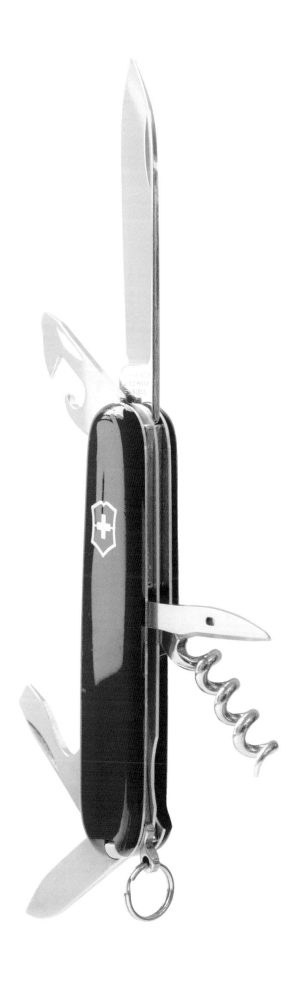

1867
Alfred Nobel

DYNAMITE

Nitroglycerine was the first new explosive to be discovered since gunpowder. Development began when Ascanio Sobrero caused an intense explosion by adding glycerine to a mixture of nitric and sulphuric acids in a laboratory at the University of Turin in 1847. Sobrero had created a liquid that was incredibly unstable and could combust at the slightest movement, creating a shock wave of 20,000 atmospheres of pressure and up to 5,000 degree temperatures, expanding at 20 times the speed of sound. Sobrero was petrified by what he had created and made no effort to commercially develop his invention.

A young Swedish chemist by the name of Alfred Nobel was the first person to have the courage to try and mass-manufacture the volatile substance in 1859, eventually employing Sobrero to work for him. Soon after starting, a huge explosion in the manufacturing plant killed five people, including his younger brother. Undeterred, Nobel conducted further experiments on a barge in the middle of a lake to make sure no one was needlessly harmed again. In 1866, a spill of potentially dangerous nitroglycerine was fortunately soaked up by a chalky packing material called *kieselguht*. The substance seemingly stabilised the nitroglycerine compound. Realising the mixture's potential, Nobel made the commercial exploitation of nitroglycerine viable by combining it with *kieselguht* and molding it into sticks for easy handling, patenting it under the name "Dynamite" in 1867. Later, he added a mercury cap to act as a detonator. The product became hugely successful and earned Nobel a fortune of around 150 million dollars, the remainder of which—after his death in 1896—was used to fund the prestigious annual Nobel Prizes.

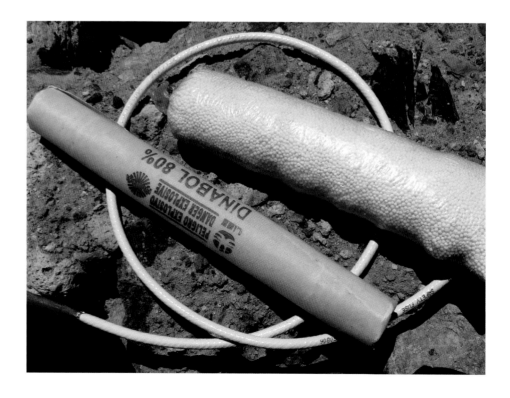

**First Century
Byzantine Empire**

HAND GRENADE

The first rudimentary grenades were used by the army of the Byzantine Empire in the first century AD; the soldiers would throw ceramic, stone or glass jars filled with 'Greek fire'; Greek fire was an incendiary weapon that would continue burning even on water. These early grenades could be easily compared to modern homemade explosive weapons used during times of civil unrest.

These small, hand-thrown, explosive bombs were used extensively around Europe during the fifteenth century; they were merely iron balls filled with gunpowder that ignited from a slow-burning wick but were particularly effective when thrown into a ditch of a fort that enemies were attacking. Grenades, so-called because of their resemblance to the fruit pomegranate, which in French is *la grenade*, became so important in warfare that during the seventeenth century European armies would train up soldiers for the specific purpose of throwing grenades: they were called "grenadiers". However, grenades and grenadiers were abandoned mid-century because new firearms utilising gunpowder had been invented that had greater accuracy and longer range, making short range and close combat artillery more or less obsolete until the early twentieth century.

The Great War lead to the re-introduction of the grenade as part of an infantryman's basic kit, proving very effective during the trench warfare that occurred. The French developed their pineapple grenade and the Germans their stick grenade, but the British infantrymen used the first really safe hand-grenade, the Mills Bomb, invented by Englishman William Mills of Birmingham. It had a central spring-loaded firing-pin and spring-loaded lever locked by a pin; once the grenade was in the air, the lever flew up and released the striker, which ignited a four-second time fuse, allowing the thrower to take cover before it exploded.

Grenades were modified so that they could be fired from guns increasing their range; they have also been modified so that they explode releasing harmful chemicals, or to produce different coloured smoke in order to create a smoke-screen for assaults, or for marking and signalling.

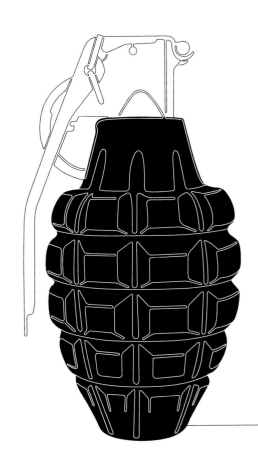

Right American Soldier throwing grenade, 1942.

1939
Robert Oppenheimer

ATOMIC BOMB

The atomic bomb is the most singularly destructive weapon ever developed, with the ability to cause instant casualties on a huge geographic scale, and causing major hereditary health repercussions for survivors.

The nuclear bomb is a device that relies on either fission or fusion reactions, or a combination of the two. Using these reactions, very small amounts of matter release vast quantities of energy; a modern one-thousand-kilogram nuclear explosive is capable of replicating the damage of over one billion kilograms of non-nuclear explosives. The atomic bomb, with its capability to devastate entire cities, is the most commonly referred to weapon of mass destruction, and is surrounded by international political agreements to keep control over their manufacture and use.

Much of the science behind the nuclear bomb comes from thinkers who were subsequently horrified at how it has manifested itself. It was Albert Einstein, famous for his discoveries in the field of quantum physics rather than the art of warfare, who triggered research into the bomb. On 2 August 1939, just before the beginning of the Second World War, he wrote to then President Franklin D Roosevelt, alongside several other scientists, of efforts in Nazi Germany to purify uranium-235, which had the potential to be used to build an atomic weapon. This was followed by The Manhattan Project in the same year; led by Robert Oppenheimer, the Project was an undercover research programme dedicated to producing the technology to make a viable atomic bomb.

In order to manufacture the device, a certain quantity of enriched uranium was needed. To gain the necessary quantity, Harold Urey and colleagues at Columbia University devised an extraction system that worked on the principle of gaseous diffusion, and Ernest Lawrence, inventor of the Cyclotron at the University of California in Berkeley, devised a process involving magnetic separation of the two uranium isotopes. Following this, a gas centrifuge was used to further remove the non-fissional uranium isotope U-238, leaving only one final stage in the process: the splitting of the atom.

Over two billion dollars was ploughed into The Manhattan Project over six years. On 16 July 1945, over the Jemez Mountains in northern New Mexico, the experimental weapon known then as "The Gadget" was successfully tested, ushering in the Atomic Age. Nothing survived beneath the huge, white mushroom cloud the device threw up.

Robert Oppenheimer made a dark statement as the bomb fulfilled what was expected of it, declaring: "I am become Death, the destroyer of worlds." Test director Ken Bainbridge was somewhat less eloquent, remarking: "Now we're all sons of bitches."

Several people involved in the project, on seeing the true horror of what they had created, signed petitions against its development as a weapon. Their protests fell on deaf ears. On 6 August 1945, the United States dropped a bomb, code-named "Little Boy", on the Japanese city of Hiroshima, and a second three days later—this a plutonium implosion-type device code-named "Fat Man"—on the city of Nagasaki. Around 120,000 people died instantly, and more fatalities followed from the long-term effects of ionizing radiation.

Opposite An aerial view of the mushroom cloud raised by the atomic bombing of Nagasaki, Japan.

Left The Operation Crossroads, Event Baker atomic bomb test explosion at Bikini Atoll, Pacific Ocean.

Middle Three cutaway views of Bushnell's American Turtle one-man submarine, 1875.

Bottom One of the German U-Boats stranded on the south coast of England after surrender.

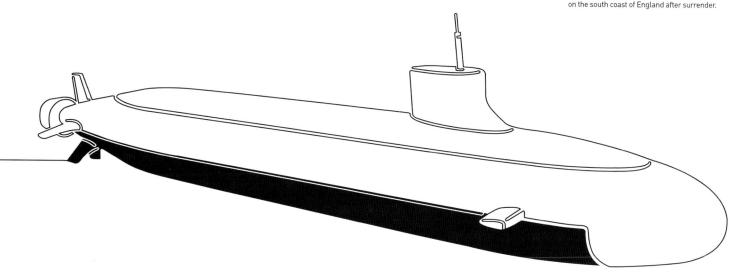

1620
Cornelius van Drebbel

SUBMARINE

The first recognisable example of the submarine concept, as with numerous modern inventions, can be seen in the work and illustrations of Leonardo da Vinci. William Bourne, a British mathematician in the sixteenth century, also drew plans for a submarine, but it was Dutch inventor Cornelius van Drebbel who attempted to actually build one, in 1620.

Van Drebbel covered a wooden rowing boat in waterproofed leather and inserted air tubes with floats to keep them above the surface at one end. The submarine was propelled by 12 oarsmen and remained successfully submerged for three hours. This early attempt at submerged travel did not come to much, and it was another century or so before the idea of the submarine was given a more practical application.

In 1776, American David Bushnell built the "Turtle", a one-man wooden submarine powered by hand-turned propellers. It was used during the American Revolution against British warships, approaching enemy ships while partially submerged to attach explosives to the hull. Though the craft itself was a success, it failed to gain notoriety due to the low standard of explosives supplied to fulfil its tasks.

Despite these early designs, the first genuine submarines were not produced until the 1890s, when two rival inventors, John P Holland and Simon Lake, simultaneously produced crafts for the United States, Russia and Japan. Their submarines used electric motors under the water, and petrol or steam engines above, also carrying motor-propelled torpedoes to afford them extra offensive potential.

In 1960, the first circumnavigation of the earth was made by the nuclear submarine USS Triton.

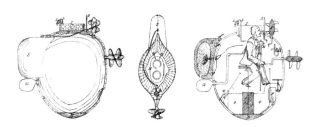

circa 1940

NIGHT VISION

The tactical advantages of engaging in combat after dark have always been well known, but it was not until well into the twentieth century that the capacity to see in the dark without also illuminating your immediate location became a possibility.

It was during the Second World War that the need for more sophisticated forms of nocturnal illumination was realised. The first, rudimentary, form of night vision involved the placing of an infrared filter over a searchlight. A viewing device then translated the infrared emissions into an electron flux that appeared on a luminescent screen. However, because these were fitted to vehicles on both sides, they could both also see each other's infrared viewers, compromising the use of the devices.

Scientists in America changed the focus of night vision technology after the Second World War, concentrating on intensifying natural light, specifically from the moon. Top secret research at Fort Belvoir, Virginia, was eventually put into practice during the Vietnam War, but the image intensifiers, known as Star-Tron scopes, were almost as heavy as the rifles on which they were mounted, and were extremely expensive. The Star-Tron scopes focused the dim night scene onto a photo-cathode inside a tube, which then multiplied the electrons from the ambient light, and focused them into a visible image. However, it would shut down if exposed to very bright light, and also emitted a high-pitched whine that exposed the user and undermined its intended stealth purpose.

The replacement of the photo-cathode with a microchannel plate (MCP), made from hollow glass tubes, eliminated these problems. The glass tubes multiplied single electrons into many, enhancing brightness. Because image intensifiers using MCPs were smaller and lighter than the older devices, they could be manufactured into binoculars and added to the helmets of soldiers. These became known as night vision goggles, and were widely used, particularly by pilots, from the 1980s.

Today, the technology of night vision is no secret, and it has practical applications even in some modern cars, to allow for safer driving at night.

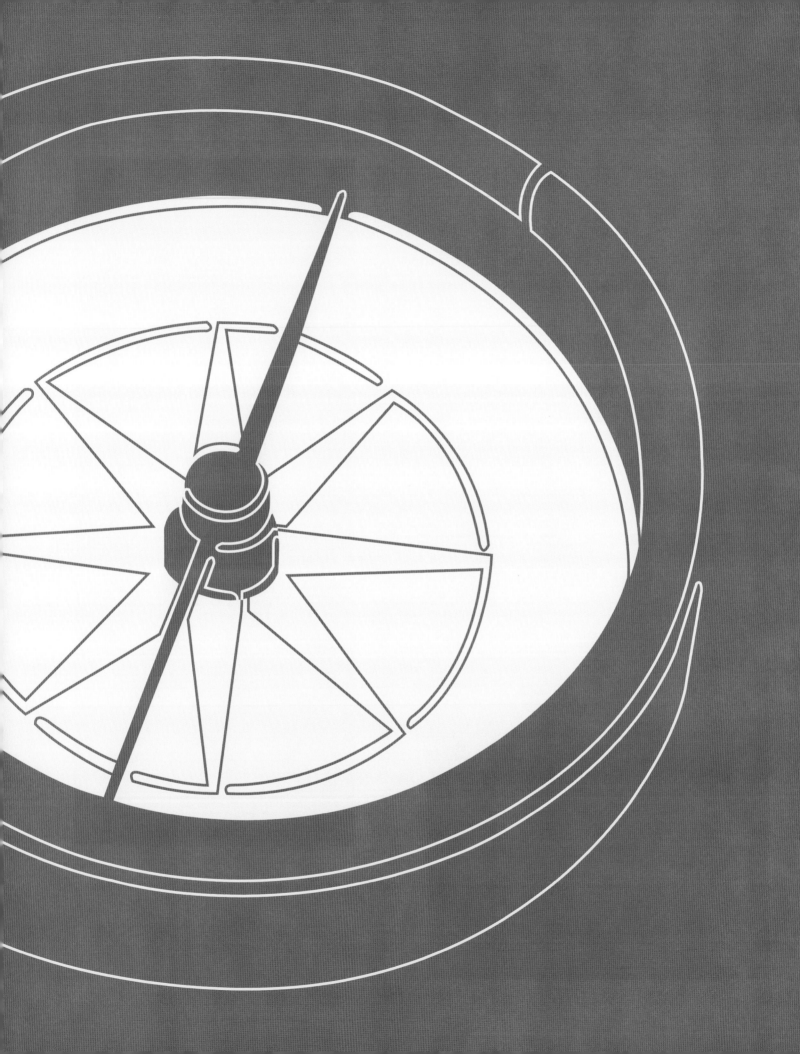

exploration

MAPS

The oldest known maps are preserved on Babylonian clay tablets from about 2300 BC, but the greatest early advances in mapping the world occurred in ancient Greece, when philosophers posited the theory of a spherical earth and revolutionised common understanding of the world. The culmination of centuries of advances in Greek and Roman thought was Ptolemy's first world map, the content of which is remarkable given that it emerged in the first century AD. Ptolemy, who wrote seminal books on mathematics, astrology and geography, was the first to provide a map that gave an impression of the world with a relative degree of accuracy, though it is far from modern standards. His authoritative guide to geography remained a valuable text until the Renaissance.

The Medieval period in Europe was distinctly less impressive, subject as it was to religious influence rather than geographic and topographic accuracy. Maps would commonly depict the world with Jerusalem at the centre and with the geographic east at the top. But while Medieval Europe floundered under religious bias, maps in Arabic countries, including those in the Mediterranean, were making more practical and representative advances.

The greatest developments in mapping began to occur in the latter stages of the Medieval period, when religious motivations gave way to the advances offered by the invention of the compass. Exploration flourished, and maps plotted by compass became increasingly accurate and exhaustive as a result. By the Renaissance, maps included compass lines, coastlines, rivers and harbours, and even general points of interest. At the same time, the invention of printing made maps much more widely available.

The first 'true' world map is generally credited to Martin Waldseemüller in 1507. His map built on an expanded Ptolemaic projection and was the first to use the name America for continent previously known as the New World.

While globes were coming into existence, the problem of the flat map of the world remained a problem. In 1569, the Flemish cartographer Gerard Mercator devised a flat map of the world that would also allow a plotted journey to appear visually as it would at sea. In other words, a journey plotted as a straight line on Mercator's map would be a straight journey in reality. The flaw in the system was that the resultant map distorted the countries it represented, but it was invaluable for mariners who could follow a route simply using his map and a compass. Eventually, as trade routes grew, the Mercator map became standard for navigation.

Although mapping continued to develop in complexity throughout the ensuing centuries, it was not until the twentieth century that maps took major new steps.

The advent of aerial photography following the First World War and satellite photography in the 1950s opened up new possibilities for mapping, a development which has only fully been realised in the twenty-first century by internet company Google. Google Earth, a three-dimensional computer model, allows the user to zoom in on any area of the world and view a satellite photograph of that area to a previously unthinkable resolution. This quiet revolution in mapping is being continued with a more aesthetic development– three-dimensional 'drive throughs' of every street in major cities of the world. An approach to mapping using geo-specific information emerged in the 1970s, known as geographic information systems (GIS). For GIS, the database, analysis, and display are physically and conceptually separate aspects of handling geographic data. Geographic information systems comprise computer hardware, software, digital data, people, organisations, and institutions for collecting, storing, analysing, and displaying georeferenced information about the earth. It is a means of mapping that Google have also attempted to incorporate into their mapping systems, using computer technology to embed geo-specific material into maps to add a new dimension of information.

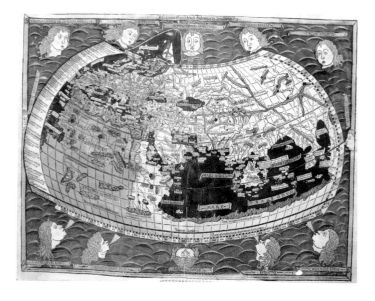

Left A fifteenth century map of the world by Nicolaus Germanius, after Ptolemy's *Geographia*.

Opposite top *Universalis Cosmographia*, a 12-panel wall map of the world drawn by German cartographer Martin Waldseemüller, published in 1507.

Opposite bottom Detail from *Universalis Cosmographia*, 1507.

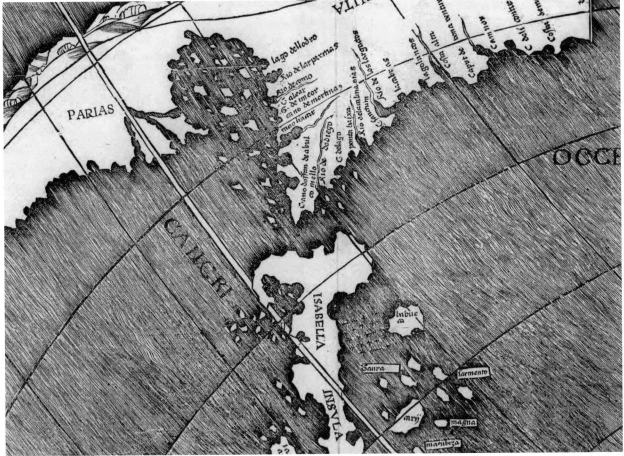

TELESCOPE

The first telescopes were made using a convex and a concave lens at opposite ends of a tube by craftsmen, many of whom were spectacle makers. For this reason it is hard to trace the telescope's origins and legitimise a claim for the first inventor of the telescope, however, the first application to patent a telescope was made by German expatriate Hans Lippershey, residing in Middelburg, Netherlands in 1609. The device that allowed for "seeing faraway things as though nearby" by a magnification of four was deemed too easy to copy so was not awarded a patent; it is this ease of assemblage that largely contributed to the vagueness of its origin.

Galileo was aware of the work of this "certain Fleming" and began work building his own, more powerful, telescope. At first he attempted to market his device for use in warfare, providing an advantage over the enemy by enhancing sight at long distances; but the military were none too carried away with the idea; in fact, a large proportion of seventeenth century society were extremely suspicious of the device as optical instruments in general were shrouded with mysticism, and therefore cynicism. Galileo therefore changed tact and promoted his device as the tool to reinvent the ancient science of astronomy. Pointing his telescope to the sky, he began to explore the heavens, becoming a propagator of the Copernican view of the world rather than the geocentric astronomy of antiquity, within the cosmological debate of the time. Galileo discovered the cratered, mountainous surface of the moon, the phases of Venus, the four satellites of Jupiter, the spots of the sun, and stars that had never been seen before. Cynics believed that it was a mere illusion, a view quite embarrassingly given fortification when the strange "companions" of Saturn that Galileo had shown his detractors supposedly disappeared. What they had actually seen was the rings of Saturn at an angle to the earth whereby they could be observed, but which latterly changed angle and were no longer observable.

Nevertheless, the telescope became one of the pioneering instruments that helped found what is commonly known as the Scientific Revolution of the seventeenth century, and became the first device to extend one of man's senses through technology. As time elapsed, progression was made via various combinations of concave and convex mirrors and lenses. Sir Isaac Newton, in 1668, was the first to produce the first practical "reflector" telescope that used a piece of diagonally placed mirror to reflect the light into a side-mounted eyepiece. The invention of the achromatic lens, designed to limit the effects of chromatic and spherical aberration, and commercialised by John Dollond in 1758, allowed telescopes to be made much shorter and significantly more practical in their use.

Many incremental developments concerning the use of materials occurred over the next century, refining the basic design of all telescopes. In 1900, George Ellery Hale built the most powerful telescope the world had ever seen for the University of Chicago with a 1.2 metre lens.

Within modern technology, radio telescopes provide the observer with the longest gaze, and can be made to pick up varied wavelengths. The American Grote Reber built the first in 1937, but it was the British who pioneered radio astronomy alongside the wartime development of radar, eventually building a giant, steerable, 76 metre dish antenna radio telescope at Jodrell bank in 1957. Two decades later 27 connected dishes known as the Very Large Array were erected in New Mexico creating a dish 17 miles in diameter; and then, in 1986, two dishes, one in Australia and one in Japan, were connected with a satellite in space creating a receiver 11,000 miles across, and capable of detecting objects whose rays had taken up to 20 billion years to reach earth.

The Hubble Space Telescope was placed into orbit by the shuttle Atlantis in 1990 and was able to detect stars shining 50 times less brightly than those seen from earth; and then 1992 saw the development of an optical telescope called the Keck-1, with a 9.5 metre reflector composed of 36 hexagonal mirrors slotted together was erected on a Hawaiian peak, which was then electronically connected to an identical Keck-2 in 1996. The universe was now observable to a range of about 14 billion light years, or 80 trillion billion miles.

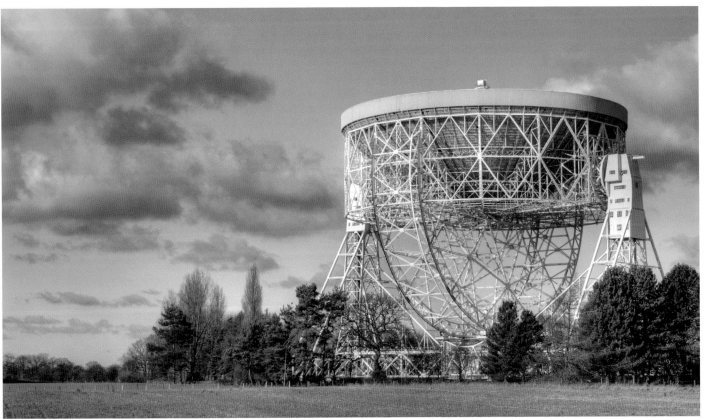

**Twelfth Century
China**

COMPASS

The invention of the compass meant that maps became relative to the traveller's actual surroundings. Before the invention of the compass, travellers would rely upon landmarks in the landscape and the use of acquired knowledge through both winds and stars; such as the reliance of the inhabitants of the northern hemisphere on the Polaris star. The creation of the compass occurred in twelfth century China and then latterly and independently in Europe, mariners unearthed a mineral which was iron-rich and naturally magnetic, aligning itself with the Polaris star. This mineral known as "lodestone"—which means "journey" in old English—became the basis of the first magnetic compass.

By 1190, Italian mariners were using needles magnetised by the lodestone and floating in bowls of water to act as ship compasses. Although these devices did not point exactly to the north and south geographic poles but rather towards the magnetic poles, they were still invaluable to mariners because of the security the device provided them, allowing them to venture further afield than ever before, in the safe knowledge that they would be able to find their way home. Accurate maps based on compass surveys began to emerge a century later and a relationship of positive feedback between the compass, expedition and cartography was ignited.

American Elmer Sperry invented the compass's modern replacement in 1908. The "gyrocompass" was based on the idea of the gyroscope invented by Leon Foucault in the previous century. With any change of direction it responds by realigning its own axis and the direction of the force in order that it points genuinely to true geographic north or south, along the earth's axis. Furthermore, the gyrocompass is immune to the effects of magnetism of iron and steel.

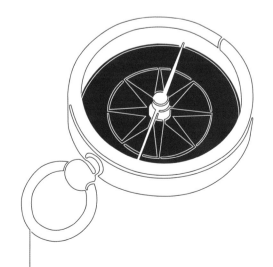

ASTROLABE

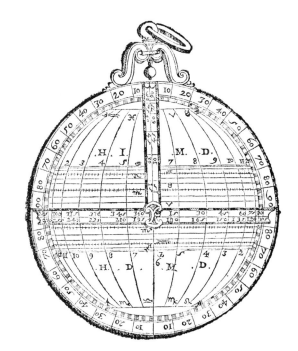

Ideas and theories that form the basis for the mechanics of an astrolabe are thought to have originated from the minds of the great pre-Christian era mathematician Apollonius and Greek astronomer Hipparchus—a man who spent his life estimating the distances of the sun and the moon from the earth; however, the first documentation of a working instrument was written by John Philoponos of Alexandria during the sixth century AD. A typical astrolabe was about six inches in diameter and made out of brass. It consisted of a disk that was marked in degrees, much like a protractor, and attached to its centre was a moveable pointer called an alidade. The astrolabe was held to the eye, whereby the baseline of the protractor was held at the same level as the horizon and the alidade was aligned with an object; from this the elevation of said object could be measured in degrees.

An astrolabe is essentially a tool of navigation and although it can be used for measuring, surveying, and triangulation, it was also commonly used by astronomers to explore the heavens for celestial mapping, locating and predicting the positions of the sun, moon, planets and stars. Early elaborate astrolabes had the positions of the planets engraved upon their disks. By finding the height of the sun and referring to right astrological tables it was possible to find out the local time, time of sunrise and time of sunset, alongside the correct latitude and direction of Mecca. This is partly the reason that the instrument became so important to the Islamic world in the eighth and ninth centuries; not only did it point to Mecca but it also allowed the user to find out the correct times for prayer, alongside its use in the study of astrology—a deeply imbedded facet of early Islamic culture.

By the fourteenth century astronomical scholarship using the astrolabe was common, but the instrument's popularity began to dwindle in the latter half of the seventeenth century after the invention of superior instruments such as pendulum clocks, telescopes and compasses. However, the astrolabe did remain the second most popular means of navigation after the compass up until the eighteenth century that saw the introduction of the sextant. The production of the astrolabe continued into the nineteenth century, especially in the Arab world, but now if they are made it is in the realm of curiosities.

Top A design of a Medieval astrolabe.

Bottom One of four extant brass astrolabes manufactured by the workshop of Georg Hartmann in Nuremberg in 1537.

1643
Evangelista Torricelli

BAROMETER

A barometer is essentially an instrument to measure atmospheric pressure. Traditional Aristotelian thought was that air had no weight whatsoever. It was Galileo who was the first to contest this view and predict the existence of atmospheric pressure caused by the force of gravity pulling weighted molecules of gas down towards the earth, but it was a student of his, Evangelista Torricelli, who invented the first device to measure this pressure.

Torricelli had been experimenting with creating a vacuum, and at the same time invented the first barometer in 1643. It was made from a four-foot long glass tube filled with mercury with a closed end at the top and an open end at the bottom inverted into a dish; some of the mercury did not escape the tube and therefore a sustained vacuum was created above the mercury: the "Torricelli Vacuum". Torricelli realised that the vacuum would support a column of mercury when it became equal to the pressure of the outside environment, he also noticed that the mercury level would change slightly in accordance with the changing pressure in the atmosphere; he wrote "we live submerged at the bottom of an ocean of elementary air, which is known by incontestable experiments to have weight".

Four years later, French philosopher and mathematician René Descartes predicted that there could be a relationship between atmospheric pressure and the weather. He applied a scale to Torricelli's tube and sent a duplicate to a colleague, Marine Mersenne, on 13 December 1647 "so that we may see if our observations agree". In effect, any fluid can be used in a Barometer, but mercury is preferred due to its weight—13.6 times heavier than water. The lighter the fluid used, the bigger the barometer would have to be, as proved by French physicist Blaise Pascal who used red wine, meaning that his barometer had to be 14 metres high. Pascal did prove, however, that air pressure fell with an increase in altitude, forcing his younger brother to climb to the summit of Puy de Dom in the Auvergne Mountains to demonstrate his theory.

The first barometer to eliminate the need of liquid by using sealed bellows—an aneroid barometer—was built by another French scientist, Lucien Vidie, in 1843. Now more portable, barometers became a common meteorological instrument in the field of research, and were used by mariners and farmers for their weather foretelling abilities. Modern day examples are controlled by electronic sensors coupled with microprocessor chips.

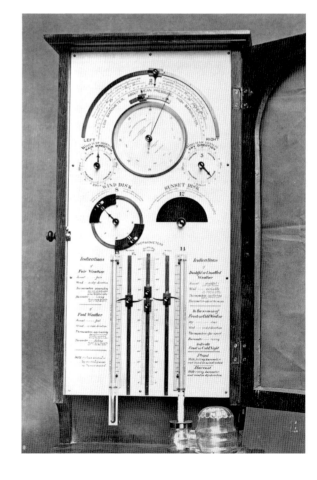

Top An antique ship barometer.

Bottom A nineteenth century English barometer.

GYROSCOPE

The gyroscope is a device comprised of a spinning disk, or rotor, whose axle is free to move at any orientation. The device is often used for navigational purposes. Given that the four points of the compass are negated in the context of space, the precession—or fixed orientation—of the internal rotor of the gyroscope can thus be used as a reference point for the measuring and maintaining of physical orientation. The fixed orientation that can be utilised derives from the event of the rotor constantly spinning. This constant circular movement means that all four points on the disk are affected if there is any change in the orientation of the gyroscope, causing it to spin on a defined axis. Modern gyroscopes are usually built to include two gimbals, rings that surround the rotor and frame. With the need to keep the inner rotors of the device stable for navigational purposes, the rings allow movement around the rotor and frame, whilst the inner components remain moving on their individually fixed axis. Gyroscopes are used in the orientation and stabilisation of aerial vehicles, large boats, and mining machinery, as well as appearing in the ubiquitous rudimentary children's toy form of the 'spinning top'.

Though German Johann Bohnenberger first wrote about the gyroscope—dubbed the "Machine"—in 1817, it was the French physicist Léon Foucault who gave the device its modern name and used it in a series of short experiments to replicate the rotation of the earth, deemed unsuccessful due to the unworkable levels of friction in the mechanism. Etymologically, the name derives from the Greek gyros, meaning circle or rotation. The emergence of electric motors in the mid-nineteenth century solved the friction problems and led to the utilisation of precession in gyrocompasses, working marine examples being developed by Hermann Anschütz-Kaempfe between 1905 and 1908. The concept was quickly picked up on by myriad manufacturers, and the Sperry Gyroscope Company soon altered the design so it could be used within military and naval crafts. Variations on these designs, still using the same basic theories, have been altered to include macro-elements such as Micro Electro-Mechanical Systems or MEMS, and are now used, as well as for the previous navigational and stabilising purposes, to record acceleration movements and data in commercial black boxes for aircraft.

As a children's toy, the gyroscope began its ascent to classic status in 1917, when the Chandler Company of Indianapolis created the "Chandler gyroscope", a miniature rotor with a pull string and pedestal which is still in production today.

Top Drawing of a gyroscope by the prolific inventor Elmer Sperry.

agriculture
and food

FOOD PRESERVATION

The basic concept behind preservation is to either kill or greatly inhibit growth of micro-organisms such as bacteria and fungi that would otherwise spoil food. To do this, the environment that the food is stored in is changed in some way. The most common form of modern food preservation is refrigeration and freezing. The cool temperature in a refrigerator inhibits bacterial action to a slow growth meaning that food which would usually perish in a matter of hours can still be edible for up to a number of weeks. Freezing, on the other hand, kills bacteria altogether. Refrigeration and freezing are popular as preservation methods because the process tends not to change the taste or texture of the food.

Canning and bottling are also popular food preservation techniques and have been used as such for over 200 years. The process involves boiling the material to kill all bacteria and then sealing the food in an airtight container so that no bacteria can enter. Boiling food can change taste and texture, but in most cases this is minimal. One method that does dramatically change both the taste and texture of the food it preserves is dehydrating. The removal of water renders the bacteria inactive, and dry foods stored in airtight packaging can stay preserved for a very long time. Common examples include vegetables, pasta, powdered milk, and soup. Freeze-drying is a method that similarly dries out food but without the unwanted side-affect of changing its fundamental characteristics; the frozen object is placed in a vacuum, which turns the water of the ice into vapour until re-opened.

One of the oldest methods of preservation involves drying out food with salt, used because it actively draws moisture out of the food packed in it. This technique was commonly used on pioneering voyages such as that of Christopher Columbus. It is such a successful technique that, if salted in cold weather before deterioration begins, meat can technically stay fresh and remain edible for years. Pickling is also a fairly antiquated method of preservation, and was traditionally used on varied foods including meats, fruits and vegetables but now is only usually used with a small number of vegetables, as well as occasionally eggs. The basic solution in which the food is preserved consists of salt and vinegar; and like salting and smoking, these add a strong, unavoidable flavour. Other more obscure preservation techniques include preservation in syrup, food irradiation, fermentation, chemical preservation, and carbonation.

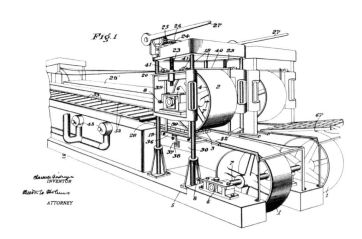

Top right Early twentieth century preserved, jarred foods.

Bottom right Patent of a refrigeration apparatus, 1930.

Left Freeze-dried ice cream.

1892
John Froehlich

TRACTOR

The internal combustion engine spawned a new era in farm machinery in the late nineteenth century. Using this new technology, Iowa blacksmith John Froehlich built the first petrol-powered tractor in 1892. Dan Albone, an English tractor design pioneer, produced the first commercially viable British equivalent in 1904, named "Iver". Turn of the century tractor designs had one major flaw, in that their intense weight would cause their wheels to sink into the often wet soil they work on. Benjamin Holt devised an ingenious solution that swapped the existing wheel designs with his newly envisioned 'caterpillar tracks', bands of linked metal that more evenly distributed the heavy weight of the vehicle.

Harry Ferguson, of the Irish Board of Agriculture, was given the task of overhauling contemporary tractor and plough designs in 1916. Prior to this date, a major fault in common models was that the caught plough or unseen dip in the earth could, given its powerful strength, cause the tractor to flip forward onto its roof, usually killing the operator. Ferguson put an end to the problem by fitting the vehicle with towing bars. Further, he made significant changes to the fundamental structure of the machine, removing the traditional large, cumbersome chassis and using the engine as the main bulk of the design, enabling the tractor to resist the huge twisting force of some heavy loads. The 1930s saw great progress in design technology. Ferguson devised the Ferguson System, which used hydraulic power to control the functions of the machines mounted on the tractor, meaning that relatively light vehicles could now operate heavy machinery. Modern day tractors use the same basic design as Ferguson's but with an extra inclusion; instead of bone-shaking steel wheels, tractors are now fitted with the instantly recognisable enormous rear wheels with rubber tyres, added for extra grip, comfort and versatility in 1932.

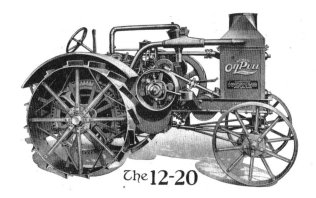

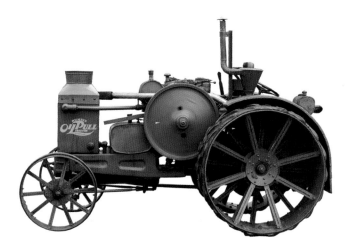

Top to bottom A two-row motor cultivator, 1918; a Whitney Tractor, 1918; a nineteenth century Persian plough, circa 1871.

Bottom left Brown's Shovel Plow advertisements, 1884.

Ancient Rome

PLOUGH

The Romans were the first to invent a machine similar to that of today's plough. The carruca was an oxen or horse-powered plough that by sixth century AD had travelled to various parts of civilised Europe from its home in the Italian Alps. A blade would dig the earth and then a mouldboard would turn over the furrow—the purpose of this process to increase the productivity of the land. The effectiveness of the plough itself was increased around the tenth century with the addition of a padded collar that allowed the beast pulling the machine to do so comfortably.

The plough became a stalwart tool in agriculture and continued to evolve providing increased facilitation for farmers. The next significant improvement of the plough's design came when steel took the place of its wooden chassis. American blacksmith John Deere was the first to create a wholly steel plough in 1837. The steel blades and mouldboards were more heavy duty than their wooden counterparts and therefore more resistant to wear-and-tear. However, Deere's steel plough was still pulled by horses or oxen and it was not until the latter part of the nineteenth century, during the industrial revolution that the agricultural world became mechanised. Larger and heavier ploughs were powered first by steam engines and then later by petrol engines. They were eventually incorporated into the design of the tractor invented by Froehlich in 1892 and were able to carry out vast amounts of work over longer periods of time.

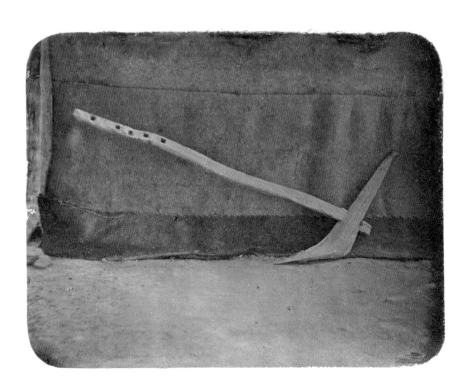

1836
Hiram Moore

COMBINE HARVESTER

Patented in 1836 by Hiram Moore, the aim of the machine was to combine the tasks of harvesting and threshing, so that crops could be cut whilst grain was simultaneously rubbed from the ears, leaving the remaining straw husk of the plant. The invention was slow to gain popularity, but by the 1870s the farmers of California were using teams of 40 horses to pull the gigantic machine, that could process one tonne of crop per hectare with its ten metre cutting width.

Modern-day combine harvesters can cost anything up to a quarter of a million pounds, are self-propelled, and commonly powered by diesel. Using rotary blades to carry out reaping, the cut crop is then deposited into a threshing cylinder by a series of chains. Grooved steel bars then separate the grain and chaff from the straw. Combine harvesters have also appropriated satellite navigation technology that allows the machine to monitor variations in weather and soil conditions as well as crop yields so that fertiliser can be applied accordingly.

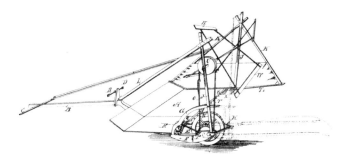

IRRIGATION

Top and bottom Ruins of ancient irrigation channel in Machu Picchu, Colombia; a nineteenth century photograph of field irrigation in Central Asia.

Bottom left A modern irrigation system.

Irrigation is the artificial process of bringing water from natural sources such as rivers and lakes, to land accustomed to long periods of time with little or no rainfall—such as deserts or cities situated at higher altitudes than a natural water source. This process involves transporting water to a level above the flood plain so that it naturally flows down towards these areas; a primitive system was first used during the late sixth century BC by the Egyptians who channelled the water from the Nile via a lever-and-bucket system. The Persians developed another system—The Persian wheel—a water wheel that collected water from a river in containers and then deposited the water into an upper course as it revolved.

The Chinese were the real pioneers of irrigation technology, devising numerous schemes from around 2400 BC. The most famous was a 1,000 mile canal that began construction in 700 BC in Hangchow and was completed by Kublai Kahn around 1280 AD in Beijing—it still remains the largest man-made water course in the world. The Romans were not far behind but they were subject to a different problem: The Tiber River was not clean enough to supply the growing city of Rome, so Roman engineers devised the now famous system of aqueducts around 312 BC. The invention of the arch allowing aqueducts to span valleys saw 11 aqueducts provide clean, fresh water to the people of Rome.

Irrigation was an important factor regarding the progress of ancient civilisations. It is now a technology indispensable to the world's crop-cultivation industry. Modern day processes not only involve traditional systems of irrigation using surface water but now also use sprinklers and micro-irrigation controlled by computers, which apply the precise amount of water in correspondence with how dry the soil is. In some parts of the world, artificial sources of water have had to be constructed in order to ensure that there are large enough reserves of fresh water for purposes such as irrigation. Engineers have built dams to create artificial lakes such as the Hoover Dam and the Franklin D Roosevelt Lake in Washington, USA, which can store, at capacity, an impressive 11.8 billion cubic metres of water.

Bottom Locally known as "bao-bao" (floating tiller), this is one of the land preparation methods used in North Cotabato, Phillipines.

1912
Arthur Clifford Howard

ROTARY TILLER

The rotary tiller is a device used to cultivate soil for growing crops. Comprising a series of turning blades, the machine works by turning up earth to increase soil fertility, and designs are made to be either self-propelled or attached to two or four wheeled tractors.

Arthur Clifford Howard invented the machine in 1912, following years of experimentation on his father's farm in New South Wales, Australia. He first began to try and devise a rotary system using a steam-powered tractor as an energy source. Early designs literally threw the already tilled soil to the side of the machine, but this problem was soon rectified via the use of L-shaped blades that turned the soil without dispersing it as animatedly. Howard had to wait until after The Great War to receive a patent for his design, comprising five rotary hoe cultivator blades, but in 1920 it finally was accepted and he set up his own company—Howard Auto Cultivators—in 1922 to produce the machines. Worldwide demand for the machines increased dramatically over a number of years, thus, with demand being hard to meet, Howard founded the company Rotary Hoes Ltd in Essex in the UK in 1938 to increase profit. Branches of the new start-up began to open worldwide, and the Howard Group exists to this day, as part of the Danish Kongskilde Industries.

PESTICIDES

A pesticide is basically any method or chemical that destroys a plant or animal competing for a human food supply. Agricultural insecticides are the most common form of pesticide and have a long history of development. Early methods of deterring insects from eating or damaging crops used by families in the Middle East included covering the plants and surrounding soil with a mixture of pyrethrum derived from the crushed petals of chrysanthemum, sulphur and arsenic; another natural example used was rotenone, which was obtained from the roots of tropical vegetables.

Large-scale application of rudimentary pesticides occurred in the late 1800s. For example, in the United States in 1877, people had to protect their potato from the destructive Colorado beetle by using chemicals that could not be absorbed by the water in the soil, and therefore would not harm crop growth. The insoluble chemicals used, in this case a compound called Paris Green, were rudimentary, and invariably were fairly ineffective.

The 1940s and 50s saw the inception of what could be termed the "pesticide era". Stronged chemicals began to be produced after the Second World War, with many products being derived from the results of contemporary chemical warfare research and appropriated for the commercial agricultural sector. DDT was the first of these, soon followed by organophosphorous compounds derived from poisonous gas such as parathion and malathion. New aircraft technology also meant that the aerial spraying of synthetic pesticides became commonplace.

Today there are over 900 types of pesticides in commercial use, the prevalence of which has increased rapidly in the last 60 years. Over 1.8 billion kilos are currently used per year around the world. However, testing has shown that traditional chemical pesticides can cause direct harm to both humans and animals. Workers in chemical fertiliser manufacturing plants and farmers who have persistent exposure to pesticides have been subject to mild pesticide poisoning, the long-term effects of which can range from memory disorders and respiratory problems to cancer and birth defects. It is also suspected that not all chemicals contained in pesticides are eradicated when crops are peeled or washed and therefore pose a significant health-risk to consumers. With media coverage of this issue heightening in recent years, people are increasingly opting for organically produced food. Pesticide use can also contaminate both water and soil, and so reduced-risk conventional chemical are being increasingly opted for, whilst biologically based pesticides such as pheromones and microbial pesticides have also become more popular.

Mid-1800s
Justus Von Lieblig

FERTILISERS

It is logical that if plants and crops absorb minerals from the soil in order to grow the soil will have to be replenished with new minerals in order to maintain the nutritional value of the soil and thus yield good crops. Traditional methods of fertilisation used cattle manure, bone meal, fishmeal, dried blood, sewage and seaweed on the land. In some cases 'green manuring' has been used, which takes place when crops are grown and then ploughed and left to decompose, depositing their nutritional value back into the soil from which they grew. The most commonly needed nutrients by plants are potassium, nitrogen, and phosphorus. Justus Von Lieblig, a German chemist, first propounded the use of these minerals as part of an inorganic fertiliser in the early part of the nineteenth century. The problem lay in the fact that it was hard to find another chemical by which to 'fix' these nutrients so that they could be used commercially. Lieblig was never able to create a suitable fertiliser himself but he did pave the way for many others. One of these was Sir John Lawes who with the help of Joseph Gilbert during the 1840s, experimented with crops and manures at his farm in Hertfordshire, England and eventually produced a practical super phosphate, using phosphates from rock and old animal dung.

The production of fertilisers using phosphorous proliferated during the latter half of the nineteenth century but similar fertilisers containing nitrogen were slow developing. The first real breakthrough came in the early 1900s via the experiments of German chemist Fritz Haber. Haber developed a technique for producing ammonia synthetically, which involved the combination of hydrogen and nitrogen with the use of a catalyst. This created a liquid compound known as ammonium nitrate, which could be used in fertilisers but originally was used during the First World War to create the first chemical weapons—his discovery became known as the Haber process and he was awarded with a Nobel Prize in 1918. The Haber process eventually meant that crop yields would be able to support the ever-increasing global population; however, the process has contributed to high levels of global pollution as only around half of the nitrogen applied to the soil is actually absorbed by the plants. Nitrates also pollute drinking water, contribute to global warming and the destruction of the ozone layer and it is now thought that nitrogen fertiliser has become the main source of pollution of the oceans. Because of this, in 2006 a new policy was introduced in both the UK and America that sees the application of fertilisers using nitrogen being regulated and controlled.

ANIMAL HUSBANDRY

Animal husbandry is the practice of breeding, raising and managing livestock and agriculture. At the end of the ice age people began to settle, forming villages and towns and moving away from a nomadic lifestyle. Subsequently animals became domesticated—first the dog around 12000 BC, then the cow whose milk and mutton became part of the human diet around 9000 BC and then the sheep, goat, and pig. Small herds of these domesticated animals would provide these communities with products such as meat, dairy, furs and leather.

Today, animals count for around 28 per cent of the world's total value of agricultural products. Modern-day methods involved in animal husbandry are usually concerned with just one type of animal—for example a swineherd cares for pigs and hogs, whereas a shepherd looks after sheep. Farming units are required to be highly efficient and often subject the animals to intense husbandry so as to produce products as quickly and efficiently as possible. This often means that large numbers of animals are confined to small pens, so that money is saved on labour and other costs such as feed. However, such methods of husbandry have been heavily condemned for being cruel to animals and producing poor quality products due to the stress and lack of exercise the animals are subjected to. These factors have headed a campaign, which has gathered popularity in the last five years or so, propagating the benefits of the production and consumption of free-range products.

Left A US animal farm during the First World War.

**800
Persia**

WINDMILL

The windmill has often been seen as one of the most groundbreaking inventions in the history of mankind. The first known windmills were the vertical axle type and were manufactured in eastern Persia, as recorded in Persian historical documents of the ninth century. In later times, a similar kind of windmill was also found in China during the thirteenth century; this type of windmill was mainly used for irrigation. The first horizontal windmills were used in the Cyclades, a cluster of islands in Greece, for the grinding of grain.

In northwestern Europe—France, England and Flanders—horizontal-axle windmills started being in use in the late twelfth century for the grinding of cereal. They were often found on top of castle towers or city walls, in Europe as well as the New World. Advancements in technology facilitated by the Industrial Revolution paved the way for steam and diesel engines but the windmill persisted well into the 1950s in certain isolated areas such the Norfolk Broads in Britain. Today such windmills are being preserved mostly for their historical significance. During the 1980s windmills started being used for commercial energy production in 'windfarms' throughout the US.

Left A wind farm at sea.

GRISTMILL AND WATER WHEEL

The development of the gristmill, or water-powered flourmill, would not have been possible without the invention of the water wheel—a superb piece of engineering that was able to harness the limitless power of flowing water to produce hydropower. Water wheels have been in use since 400 BC, their design consisting of a large wheel, which has a number of buckets or paddles along its radius that hit the water, creating force and turning the wheel. They were most probably the first instance where laborious manual work usually carried out by animals or humans was replaced by mechanics.

By 100 BC water-powered gristmills were very common throughout Egypt and much of the Mediterranean. These early gristmills would use a traditional horizontal water wheel to power the rotating stones that would grind grain into meal. The early waterwheels lay in the water horizontally and were very inefficient at transferring the power of the flowing water—they were soon replaced by vertical water wheels. This new design created two systems: the 'undershot' and the 'overshot'. The more traditional undershot system involves the use of a millpond where a stream is dammed to feed the water wheel; the wheel utilises the streams power by having its lower third submerged in the water. The overshot system came later and involved a natural stream or siphoning off of a stream or river to create a down pouring chute or small waterfall that would hit the paddles of the water wheel from a height.

There had always been a debate amongst engineers regarding whether the overshot system was better than the undershot system. Most claimed that there was no difference whatsoever, but it was British civil engineer John Smeaton who took it upon himself to find out. In 1759 Smeaton carried out various meticulous tests, his research concluded that the overshot method was twice as effective as its undershot counterpart as it had both the force of the flowing water and the force of gravity working together to create double the power. Smeaton was a pioneer in water-powered technology and created the first cast-iron water wheels. The overshot system became predominant for a short time until water-powered mills popularity dwindled during the steam-powered industrial age. However, with the advent of electricity, water was now used to produce hydroelectric power. The first ever power station in Britain was a hydroelectric station erected in Godalming in 1881.

Top A water-wheel in Rensselaer County, New York, circa 1968.

Bottom A picturesque water-wheel in Hama, Syria.

Opposite Detail of water-wheel, water-gate and forebay.

IMAGE CREDITS

Early Inventions

p. 12
Bottom: Courtesy of John Morgan

p. 13
Bottom: Courtesy of Ghewgill
Top left: Courtesy of Kio
Top right: Courtesy of A-Woody 1778

p. 15
Courtesy of James Cridland

p. 16
Top: Courtesy of A Magill

p. 19
Middle: Courtesy of Kevin Zim

p. 20
Bottom right: Courtesy of Ultramega

p. 21
Courtesy of pdxJeff

Domestic

p. 27
Courtesy of Tanais Fox

p. 29
Courtesy of Swanson TV Dinners

p. 34
Top: Courtesy of Gastev

p. 38
Bottom left: Courtesy of Dragons Fyre 1701
Bottom right: Courtesy of Eli.Pousson

p. 39
Top: Courtesy of Boliston

p. 47
Top: Courtesy of Pavelkrok

p. 54
Bottom: Courtesy of Steve Cadman

Entertainment

p. 58
Top: Courtesy of Jacob Applebaum
Middle: Courtesy of Mr T in DC
Bottom: Courtesy of wsilver

p. 62
Top: Courtesy of Ramón Paco
Middle: Courtesy of pashasha

p. 67
Courtesy of Daniel Leininger

p. 71
Bottom: Courtesy of edvvc

p. 82
Right hand side illustrations: Courtesy of Katie Fechtmann

p. 85
Left: Courtesy of editor b
Right: Courtesy of flare

p. 86
Top: Courtesy of Bain News Service/Library of Congress
Middle: Courtesy of tpower
Bottom: Courtesy of johnthescone

p. 87
Top: Courtesy of Seattle Municipal Archives
Middle: Courtesy of Lalitree Darnielle
Bottom: Courtesy of hellodrew

p. 89
Top: Courtesy of Detroit Publishing Co/Library of Congress
Bottom left: Courtesy of mikesalibaphoto
Bottom right: Courtesy of Stevage

p. 91
Both images: Courtesy of Nova Musik
www.novamusik.com

p. 94
Courtesy of Anders Sipinen

p. 96
Bottom left: Courtesy of Kevin Kohler

p. 98
Top right: Courtesy of fidgetrainbowtree

p. 102
Courtesy of CharlotteSpeaks

Communications

p. 108
Bottom: Courtesy of Bjorn Means Bear

p. 111
Top: Courtesy of Skagman

p. 112
Courtesy of Harshilshah 100

p. 113
Left: Courtesy of Zyance

p. 114
Top: Courtesy of Luc Legay

p. 115
Bottom: Courtesy of Quenerapu

p. 119
Top: Courtesy of ccgd
Middle: Courtesy of Gregory F. Maxwell

p. 120
Top: Courtesy of Javier Kohen

p. 121
Top left: Courtesy of Joe Wu
Top right: Courtesy of Joe Wu

p. 124
Top: Courtesy of Seattle Municipal Archives

p. 126
Top: Courtesy of Jimmy–Joe
Bottom: My Geo-Blog on my Garmin GPS courtesy of aburt.

p. 127
Right: Dishes Courtesy of Sanbeijl

Engineering and Transport

p. 131
Bottom Right: Courtesy of Meeshy Meesh

p. 134
Top left: Courtesy of Gamillos

p. 135
Bottom left: Courtesy of Elsie Esq.

p. 138
Top: Courtesy of Caveman
Top middle: Courtesy of Toni V

p. 139
Bottom: Courtesy of Zonnabar

p. 141
Comstock, Inc. 2000

p. 147
Courtesy of Tim Pop Up

p. 149
Right: Courtesy of tz1 1zt
Left: Courtesy of Foilman

p. 155
Courtesy of a4gpa

p. 163
Courtesy of Pikturewerk

p. 166
Courtesy of Sean McGrath

p. 168
Top: Courtesy of Crystalline Radical
Middle: Courtesy of laser 2k
Bottom: Courtesy of Karindal

p. 171
Courtesy of Studiosmith

Medicine

p. 174
Top: Courtesy of Library of Congress

p. 177
Bottom right: Courtesy of Marcel Leitnerbilder-leben

p. 178
Courtesy of WaiferX

p. 179
Courtesy of Seattle Municipal Archives

p. 180
Top: Courtesy of Tritium
Middle: Courtesy of House of Sims

p. 181
Top: Courtesy of Otis Archives 4
Bottom left: Courtesy of Internets Dairy
Bottom right: Courtesy of we-make-money-not-art

p. 183
Top: Courtesy of Stev.ie

p. 184
Top: Courtesy of Otis Archives 2
Middle: Courtesy of Otis Archives 2
Bottom: Courtesy of Stewart Dawson

p. 185
Left: Courtesy of Spider
Bottom right: Courtesy of Otis Archives 2

p. 186
Top: Courtesy of SElefant
Bottom: Courtesy of Jikatu

p. 190
Top: Courtesy of micahb37

p. 191
Top: Courtesy of Toni Barros

Warfare

p. 198
Bottom: Courtesy of Library of Congress

p. 199
Courtesy of Library of Congress

p. 205
Courtesy of Library of Congress

Exploration

p. 215
Bottom: Mike Peel, 2007, www.mikepeel.net

p. 217
Bottom: Courtesy of Sage Ross

p. 218
Bottom: Courtesy Library of Congress

Agriculture and Food

p. 222
Top: Courtesy of Library of Congress
Bottom: Courtesy of Gustav H

p. 225
Bottom: Courtesy of dok1

p. 226
Bottom right: Courtesy of Library of Congress

p. 228
Top: Courtesy of Eduardo Z
Bottom right: Courtesy of Library of Congress
Bottom left: Courtesy of Chris Happel

p. 229
Top: Courtesy of Orphan Jones
Bottom: Courtesy of Keith Bacongco

p. 230
Top: Courtesy of Paul L. Nettles
Middle: Courtesy of Dive Master King

p. 231
Middle: Courtesy of Boboroshi

p. 232
Top: Courtesy of BONGURI
Bottom: Courtesy of Library of Congress

p. 234
Both courtesy of Library of Congress

p. 235
Courtesy of Library of Congress

All in-house photography:
Alex Wright for Black Dog Publishing.

COLOPHON

Edited by Duncan McCorquodale, Phoebe Adler, Tom Howells
and Paul Sloman. Texts by Louis Hill, for Black Dog Publishing.

Designed by Alex Wright with support from Matt Bucknall
and Katie Fechtmann at Black Dog Publishing.
Cover design and interior illustrations by Live Bergitte Molvær
at Black Dog Publishing.

Black Dog Publishing Limited
10a Acton Street
London WC1X 9NG
United Kingdom

Tel: +44 (0)20 7713 5097
Fax: +44 (0)20 7713 8682
info@blackdogonline.com
www.blackdogonline.com

British Library Cataloguing-in-Publication Data.
A CIP record for this book is available from the British Library.

ISBN 978 1 906155 67 4

Black Dog Publishing Limited, London, UK, is an environmentally
responsible company. *Inventors and Inventions* is printed on an
FSC certified paper.

architecture art design
fashion history photography
theory and things

www.blackdogonline.com